Power FETs and Their Applications

Edwin S. Oxner

Staff Engineer
Siliconix incorporated

PRENTICE-HALL, INC., *Englewood Cliffs, N.J.* *07632*

Library of Congress Cataloging in Publication Data

Oxner, Edwin S.
  Power FETS and their applications.

  Includes bibliographies and index.
  1. Field-effect transistors.   2. Power transistors.
I. Title.
TK7871.95.96   621.3815'284   81-15713
ISBN 0-13-686923-8   AACR2

Editorial production/supervision: Ellen Denning
Interior design: Ellen Denning
Manufacturing buyer: Gordon Osbourne
Cover design: Edsal Enterprises

Printed in the United States of America

10 9 8 7 6 5 4 3 2

ISBN   0-13-686923-8

PRENTICE-HALL INTERNATIONAL, INC., *London*
PRENTICE-HALL OF AUSTRALIA PTY. LIMITED, *Sydney*
PRENTICE-HALL OF CANADA, LTD., *Toronto*
PRENTICE-HALL OF INDIA PRIVATE LIMITED, *New Delhi*
PRENTICE-HALL OF JAPAN, INC., *Tokyo*
PRENTICE-HALL OF SOUTHEAST ASIA PTE. LTD., *Singapore*
WHITEHALL BOOKS LIMITED, WELLINGTON, *New Zealand*

Contents

Preface

The introduction of the first commercially available power FET in the fall of 1976 caused such a stir that before the decade closed, semiconductor manufacturers throughout the world considered it prudent to follow suit. It now seems that every month we are surprised by the introduction of a new power FET ready to dazzle us with benefits and features, often trying, sometimes successfully, to outdo its predecessors.

Yet amid all the fanfare of a new product announcement, we find ourselves possibly a bit overwhelmed and perhaps groping for a clearer understanding and appreciation as to the usefulness of these new power FETs. Data sheets and salespeople expound benefits and features, but the questions we continue to hear remind us that no product is truly successfully introduced until it finds its way into practical applications, where it can go on to prove itself in performance.

This book is dedicated to answering these questions: "What is a power FET?" "Why do I need power FETs?" "Where can I use the power FET?" "How do I use power FETs?"

We begin by answering the first question and comparing FETs with several power semiconductors. A brief history of the evolution of power FET development closes the first chapter. The next two chapters begin by comparing the numerous types of power FETs that are available: their features, their performance, and their characteristics. These chapters close with a review of how the various power FETs are constructed, which, hopefully, helps us better to understand their operation and assists us in choosing the right FET for the job.

Chapter 4 is of special importance to the engineer interested in modeling his circuit. Although data sheets abound, none provide the designer with definitive modeling information. Since the designer may be familiar with bipolar transistors, this chapter will help provide meaning to what may be new terms.

The remainder of the book we dedicate to applications ranging across major areas of interest to the widest possible audience. Each chapter follows a common pattern, first identifying the special application, the problems, and the limitations encountered when not using power FETs. Block diagrams and simple circuits with explanation follows. The chapter closes with a summation of the benefits and potential problems associated with power FETs in that application. Following each chapter we have a comprehensive list of References providing the reader with collateral reading. In the last chapter we briefly review how we should select the right FET for the right application.

Many individuals and companies supported me in this work. Those individuals whose names come to mind are Ranier Zuleeg, who provided me with a wealth of data on the Gridistor, the MUCH-FET (which he invented), and the static induction transistor. My reviewers, Rudy Severns, consultant, and Jerry Willard of Tektronix, offered many helpful suggestions that I am sure have materially improved the text and, in particular, that portion within Chapter 5, where Rudy and International Rectifier graciously provided me with heat-sink information. I have tried, wherever possible, to give recognition of the companies that graciously provided permission to use their material.

I cannot close without extending my warmest appreciation to my wife Carol and to my children, who provided continual encouragement. Finally, I would never have written this book if my friend, Doug DeMaw of the American Radio Relay League, Inc., had not asked David Boelio of Prentice-Hall, Inc., to seek me out. To them I owe special thanks for making me an author.

E.S.O.

one

Introduction

1.1 FIELD-EFFECT TRANSISTOR BASICS

There are two basic field-effect transistors, which we shall hereafter in this book call *FETs*. These are the junction FET, or *J-FET*, and the *metal-oxide semiconductor* FET, more commonly called the *MOSFET*. Continuing our definition we can further distinguish these types as *n*-type and *p*-type, much like the well-known *npn* and *pnp* bipolar transistors, in which the letters identify the substrate dopant.

FETs can be made from several basic semiconductor elements, which have peculiar relationships when we view the periodic table of elements. The basic elements comprising both J-FETs and MOSFETs fall into group IV and the doping elements that determine whether the transistor is *n*-type (or for a bipolar transistor *npn*) or *p*-type (again, for the bipolar transistor *pnp*) fall into groups III and V. Since it is not our purpose to study the physics of semiconductors, the reader is invited to research the literature offered in the References.

Whereas we know that bipolar transistors have been constructed from both germanium and silicon, the preponderance of industrial power transistors have been limited to silicon, and for good reason. Silicon and germanium technology are both well understood, but silicon offers us two advantages that have made it more appealing:

higher operating temperatures and lower leakage. These two benefits alone have allowed us to witness the slow demise of germanium in power bipolar transistors and the total rejection of germanium in power FETs. One reason for rejecting germanium is that a low operating temperature mandates lower thermal dissipation, which, in turn, dictates lower operating power. As we shall discover in later chapters, this becomes an unacceptable boundary that we cannot negotiate in the development of power FETs.

There has been considerable excitement in the design of super-high-performance *gallium-arsenide* power FETs for application in the microwave frequencies—which is quite another subject and one we are not prepared to study in this book. Our focus will be on silicon power FETs to the exclusion of all others.

1.1.1. The Difference between J-FETs and MOSFETs

When we view the family tree in Fig. 1-1, we see two major branches, one of J-FETs, the other of MOSFETs. As we continue in our study of the power FET we will discover that both the J-FET and the MOSFET play contributing roles. Consequently, we will pause for a moment in our study of the power FET to identify the differences between J-FETs and MOSFETs and to recognize the salient features that distinguish them from one another.

The J-FET and the MOSFET have fundamental fabrication differences which are quite evident in Fig. 1-2. Here we see that the J-FET has a diffused or implanted *gate electrode* in a bar of semiconductor material. The MOSFET, on the other hand, has a gate separated by an insulating layer, usually an oxide. Without our dwelling on detail, we can immediately perceive a major difference in operation between the J-FET and the MOSFET. For the J-FET, if we connect the gate electrode to the *source electrode*, the applied voltage will appear across the gate-to-drain region. Since the gate is reverse-biased in this illustration, we find that a *depletion region* exists between the gate and drain which sweeps away all the electrons. Since current flow is in reality electron flow, we see that as the depletion region expands, the current will fall. Conversely, we recognize that there is also a limit to the amount of current that can be driven through the channel. For the J-FET this limiting current is known as the *saturation current*.

For the MOSFET we have two options, one wherein the diffused channel extends between source and drain, the other with no apparent channel [as in Fig. 1-2(b)]. Both have a gate electrode separated from the bulk semiconductor material by an insulating medium which we previously identified as an oxide layer. For either, MOSFET performance is controlled by the gate which when a potential is applied acts

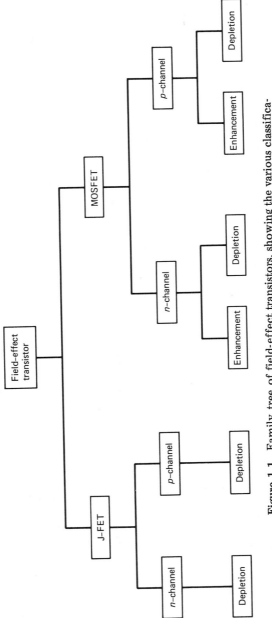

Figure 1.1 Family tree of field-effect transistors, showing the various classifications possible.

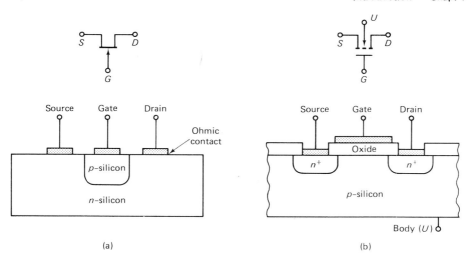

Figure 1.2 Comparison of cross-sectional views between a J-FET (a) and a MOSFET (b).

very much like a charged capacitor, the other plate being the semiconductor material. If we wish to inhibit the current flow in the former MOSFET (where we see the channel extending between the drain and the source) we simply impress a potential of opposite polarity upon the gate from that impressed on the drain. The gate, acting as a charged capacitor, generates a field in the channel that depletes the electrons and, much like the J-FET, current flow is restricted. However, since the gate electrode is not a *p-n* junction as is the J-FET gate, one can reverse-bias the gate of a MOSFET without harm. By so doing we can enhance the electron flow through the channel, causing an appreciable increase in current flow.

The latter MOSFET (with no apparent channel) is quite obviously in a nonconducting state without the presence of a gate potential. To enhance current flow we must place a potential of like charge on the gate, which in turn places an opposite charge on the semiconductor material. We now witness a new phenomenon: this charge *inverts* the semiconductor material, forming a channel for current flow. For our illustration in Fig. 1-3, let us assume that the bulk semiconductor material is *p*-type, the source and drain electrodes are *n*-type, and the supply voltage is positive. Placing a positive potential on the gate electrode attracts electrons in the semiconductor material under the gate. An overabundance of electrons in formerly *p*-type semiconductor causes what is known as *inversion*. The *p*-type becomes *n*-type and current begins to flow.

As we conclude this hasty review of FETs, we can now identify the fundamental performance of a FET by its name. For example, the

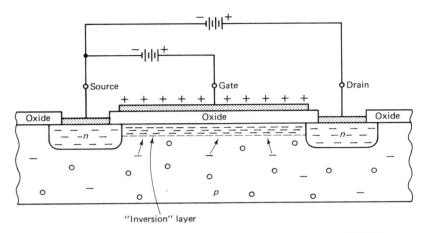

Figure 1.3 Inversion of an *n*-channel, enhancement-mode MOSFET. A positive gate bias attracts opposite charges under the oxide, which effectively makes the *p*-doped substrate predominantly *n*-enhanced, thus tying the source and drain together.

J-FET is always a depletion-mode FET, whereas the MOSFET can be either depletion or enhancement, or in some cases both.

1.2 WHY POWER FETS?

Power field-effect transistors are fast overtaking the ubiquitous power bipolar transistor in many applications where only a few years ago the power bipolar transistor was dominant. Why this sudden interest in power FETs? What has turned our attention to consider the power FET? In the next few sections we attempt to find answers to our questions.

Although there are many types of power FETs, which we will explore in depth in Chaps. 2 and 3, all FETs, whether large or small, whether they handle amperes or microamperes, have a basic similarity: they are *majority-carrier* bulk semiconductors.

1.2.1 Comparing the Power FET to the Bipolar Transistor

Bipolar transistors, on the other hand, also have majority carriers in addition to minority carriers. To appreciate the benefits offered by FETs let us begin by examining the structural differences between the junction FET (J-FET) and the bipolar transistor. In Fig. 1-4, we see *p-n* junctions common to both. For the J-FET this *p-n* junction appears in shunt, whereas for the bipolar transistor they are back to back

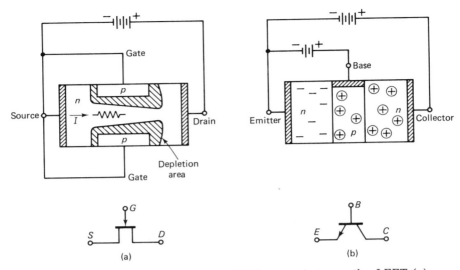

Figure 1.4 Fundamental structural differences between the J-FET (a) and the bipolar transistor (b).

and in series, hence the acronym *BJT* for the bipolar junction transistor. A simple equivalent circuit of an *n*-channel J-FET and an *npn* BJT is shown in Fig. 1–5, which will help us with our questions.

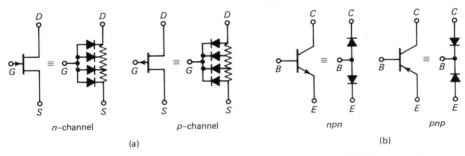

Figure 1.5 Symbols for and simple equivalent circuits of J-FETs, *n* and *p* channel (a), and bipolar transistors, *npn* and *pnp* (b).

Recognizing the *p-n* junction as a basic diode, we can identify at once the fundamental difference between the J-FET and the BJT. If we place a potential across both, current will flow through the J-FET but will be stopped by the back-to-back (*np-pn*) diode pair intrinsic to the BJT. Since it is not our intent to analyze the bipolar junction transistor, BJT (we are using the bipolar *junction* transistor because we are comparing performance to a *junction* FET), it is sufficient to say that it operates by the mechanics of *injection*, *diffusion*, and *collection*.

In this comparison we are interested principally in the mechanics of injection and its effect upon performance. In our example, shown in Fig. 1-5, it is necessary to inject positive carriers into the base region to begin current flow between the collector and emitter. The injection of carriers into the base are called *minority carriers* and the collector to emitter current flow is called *majority-carrier* current flow.

1.2.1.1 Benefits of FETs compared to bipolar transistors. The first major benefit offered by the FET in contrast to the BJT focuses upon two effects relative to minority-carrier injection. Remembering that a FET does not have minority carriers, we can deduce the first effect from examining Fig. 1-5(b). The injected carriers (positive for an *npn*, negative for a *pnp)* forward-bias the base-emitter diode, which results in both collector current (majority-carrier) flow and base-to-emitter current (minority-carrier) flow. Using the BJT as an analog switch, the output signal when ON would have an offset caused by the minority-carrier base-emitter current superimposed upon the majority carrier collector-emitter current.

For the FET, majority-carrier current is controlled by a gate potential rather than by minority-carrier injection. When used as an analog switch, we see no offset. Figure 1-6 clearly shows the comparison.

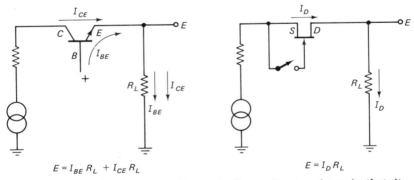

$$E = I_{BE} R_L + I_{CE} R_L \qquad\qquad E = I_D R_L$$

Figure 1.6 An advantage of a majority-carrier transistor is that it provides no offset voltage.

The second benefit is the absence of the well-known minority-carrier storage time characteristic of all bipolar transistors, which severely limits their switching speed. This effect becomes increasingly more pronounced in the bipolar transistor as the current-handling capacity increases. The FET, on the other hand, not having minority carriers, does not suffer from this effect, so we are not limited in switching speed by this effect.

This benefit hinges upon the concept that a FET is a *bulk* semiconductor. Examining Fig. 1-4, we see that the current flow from drain to source is impeded only by the bulk resistivity of the semiconductor medium and the ohmic contacts to that medium. On the other hand, we have previously seen that the collector-to-emitter majority-carrier current flow in the BJT begins with the injection of minority carriers, which excites what is known as *transistor action*.

1.2.1.2. Temperature effects. In linear applications such as in amplifier circuits, an increasingly heavier current flow (of majority carriers) will result in a progressively increasing temperature rise in the semiconductor. Bulk silicon semiconductor resistivity will change at a rate of approximately 0.6% per °C.

If we impress a voltage across a FET, as the temperature rises the majority-carrier current will drop in accordance with Ohm's law. We may consider the FET to be *thermally degenerate*.

In comparison, the transistor action of the BJT is more complex, giving rise to increasing activity with increasing temperature. This suggests that the bipolar transistor, unlike the FET, is *thermally regenerate*. Simply stated, a bipolar transistor left without protective bias may break into what is known as *thermal runaway*, which could easily lead to catastrophic destruction.

Because the FET is thermally degenerate, we can identify additional benefits that are not available to the bipolar transistor. If we need to control currents in excess of what a single FET can manage, we simply parallel a sufficient number of FETs until our current requirements are satisfied. Unlike the bipolar transistor, no current-equalizing procedures are required to prevent current hogging and possible thermal runaway. The degenerate characteristic acts to equalize any intrinsic imbalance between individual parallel FETs.

At this point it should be noted that since FETs are majority-carrier semiconductors, paralleling does not necessarily increase their drive requirements. Fan-out from logic drivers can be nearly limitless. We treat this more fully in Chap. 9 when we discuss using the power FET as a switch. We will also discover a major benefit that results from this unique thermal property that simplifies the construction of power FETs as we review various fabrication techniques in Chap. 3.

A final thermally related benefit, which we describe in greater detail in Chap. 4 when we model the power FET as an amplifier, is the freedom of secondary breakdown. To fully appreciate this benefit, in Fig. 1-7 we can compare the *safe operating area* (SOA) of a power FET with that of a comparably rated power bipolar transistor.

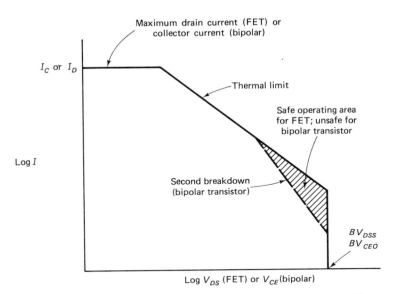

Figure 1.7 Comparison of the safe operating area (SOA) of a Power FET and a bipolar transistor.

1.2.2   The Structure of the SCR Compared to the Power FET

The SCR is one of a family of *pnpn* devices. If we examine the schematic and simplified model in Fig. 1-8, we note that structurally the SCR belongs to the family of bipolar transistors, or to be more correct, we should say that it belongs to the family of thyristors. Operationally, it is a rectifier capable of passing current in only one direction. Second, like the FET, it has a gate electrode. As a rectifier it

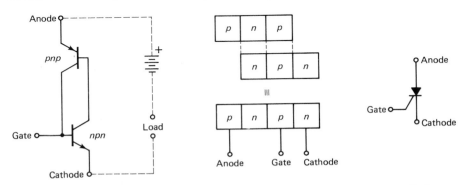

Figure 1.8 Electrical schematic and model of a silicon-controlled rectifier (SCR).

is unique, for without a burst of minority carriers injected into the gate the SCR will not conduct. The gate, as viewed in the simplified model, is quite obviously analogous to the base of a bipolar transistor but with the unusual limitation that allows it to turn the SCR ON but not to turn the SCR OFF!

1.2.2.1. The shortcomings of the SCR compared to the power FET. Although the SCR can switch megawatts at high standoff voltages, nonetheless, it exhibits one severe shortcoming: turn-off time. To turn the SCR OFF we must interrupt or reduce the forward conduction current long enough to allow the excess majority carriers to either recombine (hole-electron pairs) or to bleed off. Until we return the structure to its blocking mode, the SCR will remain ON. In ac circuits commutation of the sinusoidal wave reverse-biases the SCR, allowing it to return to its blocking mode. To maintain action the gate must trigger after each half-cycle. In dc circuits we must resort to external circuits to reduce the anode current for the SCR to regain its blocking state. Since turn-off relies upon carrier recombination (much like the bipolar transistor), either method is considerably slower than the switching time of the power FET. A medium-sized SCR may take a few microseconds to turn off completely, whereas we can measure the power FET in nanoseconds.

1.2.3 Comparing the Power FET to the Triac

The Triac is closely related to the SCR, with the additional benefit of being able to pass current in either direction. Like the SCR we see a three-terminal device shown schematically in Fig. 1–9. Unlike the SCR, the Triac does not limit us to conduction of less than $180°$ when used to control ac circuits. However, like the SCR, the Triac suffers from slow switching times and, in addition, we must guard against misfirings

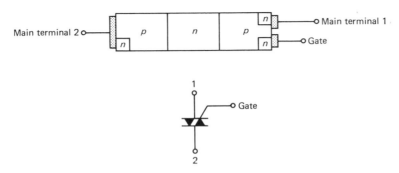

Figure 1.9   Electrical schematic and model of a Triac.

when high-speed transients are present on the ac line. In practice we might see transient suppressors shunting the Triac, which helps to prevent false triggering that might result from excessively high reactive loads.

As we leave the subject of thyristors we must remember that they are quite sensitive to temperature rise, so adequate heat dissipation is a necessity.

In Chap. 5 we explore in some detail the use of power FETs in power supplies that heretofore were the domain of both SCRs and Triacs.

1.2.4 Comparing the Power FET to the Darlington Pair

If we were challenged to simulate the electrical characteristics of the power FET using bipolar transistors, possibly the closest analogy that we could use would be the arrangement shown in Fig. 1–10, called the *Darlington pair*. This double emitter follower offers us both the high input resistance and high gain that so typifies the power FET. By choosing suitable bipolar transistors and biasing we can achieve current gains (*beta*) of thousands and input resistance in the thousands of ohms.

Although we find Darlington pairs used in many diverse applications, they nevertheless have three shortcomings that will eventually have us see them replaced with power FETs.

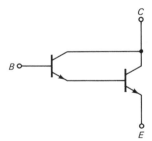

Figure 1.10 Schematic of a typical Darlington pair.

1.2.4.1 Shortcomings of the Darlington pair compared to the power FET. Having reviewed the shortcomings of the power bipolar transistor in Sec. 1.2.1, we can anticipate that the benefits offered by power FETs in comparison with the Darlington pair should focus mainly on the switching and thermal characteristics of the latter. If minority-carrier storage time was a problem for a single bipolar transistor, it is easy for us to appreciate the compounding of the problem with two bipolar transistors in cascade. Turn-OFF time for a Darlington pair is excruciatingly long in comparison to a power FET, even when we use especially constructed monolithic Darlington pairs designed for "fast" turn-OFF. Where turn-OFF for the typical power FET remains in the low-nanosecond region, for the Darlington pair it may well remain in the low-microsecond range!

The thermal problem of the Darlington pair is a variable problem depending to some extent upon our demands for current gain, voltage rating, and drive requirements. More often than not we find that the Darlington pair has increased dissipation simply because of the second bipolar transistor's inability to saturate fully when in the ON state. Regardless of our choice, if we choose to operate at ambient temperatures much above $140°C$, we are simply asking for trouble. Thermal runaway can literally wipe us out.

If we try to emulate the power FET's extraordinarily high gain, we then discover that the Darlington pair's saturation voltage (V_{sat}) becomes awkwardly high unless special biasing provisions are made. To be sure, a high saturation voltage was the plague of power FETs when they were first introduced in the late 1970s, but within a few years the technology had improved so that today we are able to select power FETs with saturation voltages comparable to the best of bipolar transistors and in some cases even lower!

1.2.5 Reviewing the Benefits of the Power FET

We have concluded our comparison of many popular power semiconductors and hopefully we have found the answer to our question. Why is the power FET so interesting? If, indeed, it exhibits the characteristics that we have identified and if its benefits are as we have been led to understand, our question is answered. But before we close this section, let us review the benefits of the power FET.

At this point it is premature for us to differentiate between the many different types of power FETs and to identify their own specific characteristics. We will have that opportunity as we finish Chap. 2. For now, let us concentrate on a review of what we have covered up to this point. For convenience we will identify the features of power FETs followed by the benefits that these features offer.

In Chap. 2 we will begin to examine many types of power FETs and compare their various features. As we do we will be able not only to add to this list of features and benefits, but better yet, we will develop a keen perspective which will help us to identify the optimum application for each type. We will see some power FETs suited for motor control, others that show exciting potential in high-frequency applications. But, best of all, we will develop an understanding that will give us the wisdom of choice.

With benefits such as these and with a promise of more to come in

Features	Benefits
No minority carriers	High gate input resistance Very high gain (beta) Very high speed switching Nearly unlimited fan-out (if we disregard switching speed)
Thermally degenerate	No current hogging Greatly inhibited thermal runaway Greatly inhibited secondary breakdown Improved safe operating area Accepts high inrush drain current

Chap. 2, what happened that delayed the arrival of the power FET? The bipolar transistor was invented in the late 1940s and power bipolar transistors followed immediately. MOSFETs, on the other hand, were known in the very early 1960s and it took nearly two decades before power FETs appeared. Before we close this chapter we will digress into how the power FET technology evolved.

1.3 THE DEVELOPMENT OF THE POWER FET

History is subjective. Its meandering path frequently appears to be following the trail hewn by the historian. We will try a more direct approach by limiting this survey to those silicon power FETs that were seen, studied, and reported, hoping to forsake those that, to the best of our knowledge, never developed beyond simply an idea. People are a natural part of our history, so if some names appear to have been forgotten, please excuse us for the omission.

1.3.1 In the Beginning . . .

It was in 1930 when the field-effect principle was first disclosed in a U.S. patent by a Julius Lilienfeld, a former professor of physics at the University of Leipzig who had recently immigrated to the United States.

Nearly two decades later, in 1948, Shockley and Pearson tried fabricating a rudimentary FET using evaporated layers of germanium on dielectric. However, it was not until Bardeen theorized on the surface state phenomenon and Shockley published his theoretical analysis of the unipolar field-effect transistor that we saw any material advance in the fledgling technology.

Another hiatus brings us to 1964, when two significant papers were published, one by Zuleeg, the other by a French physicist, Teszner. It was in these papers that FETs were first considered suitable for handling power. Were we to review these papers we would see a remarkable similarity, for both purported to have solved two mutually exclusive problems: the "high-frequency" problem and means to achieve high saturation currents. It was the latter solution that opened the door for the development of the power FET.

Before we continue we should first understand the "high-frequency" problem. It was customary at that time to achieve higher gain and higher current-handling capacity simply by increasing the physical size of the FET geometry. That, we can see, was self-defeating simply because increasing the size also increased the parasitic capacitances. Additionally, an increase in the channel resistance occurred. Together they spelt doom for high-frequency performance. The higher parasitic capacitance reduced the frequency response while the increased channel resistance lowered the forward transconductance and limited the current.

It appeared that both Zuleeg and Teszner had resolved the problem. Zuleeg called his device a *multichannel field-effect transistor*, which he nicknamed the MUCH-FET. Teszner, in his coauthored paper, called his the *gridistor*, as he felt that it closely resembled the triode vacuum tube except, of course, it was solid state. Actually, both Zuleeg and Teszner had realized an equivalent vacuum-tube triode called the SCL (*space-charge-limited*) triode. Zuleeg, however, had made another major contribution, for his MUCH-FET was fabricated *vertically*, as is practically every power FET that we find in the market today.

During this activity the Japanese were quietly active and a pair of researchers at Tohoku University, Nishizawa and Watanabe, in 1950 patented in Japan what they called an analog transistor. It was a FET.

About the time that patents were being issued to both Zuleeg and Teszner for their inventions, the U.S. Army and later the Department of the Navy released a research contract to RCA to develop a high-frequency power FET. As it turned out when the final report, underwritten by the Army, was issued in 1968, it did not use either Zuleeg's nor Teszner's ideas. Had they used the solutions proposed by these two researchers, the money spent might have resulted in a more worthwhile contribution.

Simultaneous with the work of Zuleeg and Teszner, the Japanese were developing their analog transistor field-effect transistor, a work that had remarkable parallels to the work of Zuleeg. This analog transistor was later to be named the SIT (*static induction transistor*), capable of remarkable power-handling abilities at high frequencies.

Concurrent with this work by the Japanese research team of Nishizawa and Watanabe, the Japanese Electrotechnical Laboratory, in

1969, reported what we were able to read as a major technological breakthrough. This we discovered was what has now become the well-known vertical V-groove MOSFET structure that no longer relied upon the limited dimensional accuracy of the photolithographical process.

As the decade of the 1970s closed, power FET development had spread worldwide with each year announcing a new technology. The VMOSFET was the first commercially available power FET we saw followed by a vertical DMOS (*double-diffused MOS*), the most note-worthy being the HEXFET[1]. In this decade the Japanese introduced their SIT, where application was soon to settle in high-fidelity audio power amplifiers.

Even the Soviets announced a "revolutionary" power FET pur-ported capable of several watts output to 100 MHz (their KП901A series).

Figure 1–11 displays several of the early power FETs that we have read about in the last few pages.

Figure 1.11   Some of the power FETs available today.

[1] HEXFET is the trademark of International Rectifier, Inc., Semiconductor Division, El Segundo, Calif.

REFERENCES

HUNTER, LLOYD P., *Handbook of Semiconductor Electronics*. New York: McGraw-Hill Book Company, 1970.

MALONEY, TIMOTHY J., *Industrial Solid-State Electronics*. Englewood Cliffs, N.J.: Prentice-Hall, Inc., 1979.

SHOCKLEY, W., "A Unipolar 'Field-Effect' Transistor," *Proc. IRE*, 40 (November 1952), 1365–76.

TESZNER, S., and R. GIQUEL, "Gridistor—A New Field-Effect Device," *Proc. IEEE*, 52 (1964), 1502–13.

ZULEEG, R., "Multi-Channel Field-Effect Transistor, Theory and Experiment," *Solid-State Electronics*, 10 (1967), 559–76.

two

Types of Power FETs

2.1 INTRODUCTION

In Chap. 1 we identified the two basic FET structures: the J-FET and the MOSFET. From our examination of Fig. 1–1, we concluded that the J-FET, whether it is constructed as an n-channel or as a p-channel device, operates in what we commonly call the depletion mode. The MOSFET, on the other hand, again irrespective of its polarity (n-channel or p-channel), can be constructed so as to operate in either the depletion mode or in the enhancement mode.

Perhaps at this point it would be well for us to review briefly the important *operational* differences between depletion-mode and enhancement-mode FETs. Since all power FETs must, by definition, be either depletion mode or enhancement mode, a basic understanding of the operational principles is endemic to the understanding and use of power FETs.

2.2 OPERATION: DEPLETION MODE

Since we have previously associated the J-FET as a depletion-mode device, we shall use the J-FET as our model and examine its operation.

A depletion-mode FET has two operating regions: the first is often called the *triode* region and the second the *pentode* region. We can more closely identify these operating regions by examining the family of output characteristics in Fig. 2-1. In the region below the

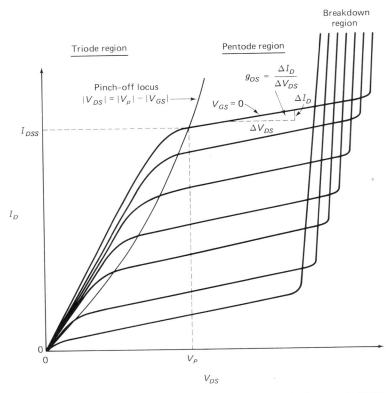

Figure 2.1 Typical output characteristics of a depletion-mode FET.

pinch-off locus we have the triode region, where I_D is governed by V_{DS} (and to a somewhat lesser extent by V_{GS}). Above the pinch-off locus we see that the characteristics resemble the pentode or *saturated* region, where I_D is controlled exclusively by V_{GS}. These two regions are important. We operate the J-FET in the triode region when we use it as a switch and in the pentode region as a voltage amplifier.

We can visualize the J-FET operating both in the triode and pentode regions by referring to Fig. 2-2. In Fig. 2-2(a), the drain-source voltage, V_{DS}, is less than the pinch-off voltage, V_P, and the resulting depletion area constricts the channel so as to reduce the flow of majority carriers according to Ohm's law. This we call the triode region or *resistance* region. In Fig. 2-2(b) we have increased V_{DS} to a potential greater than V_P, causing the depletion to so constrict the channel

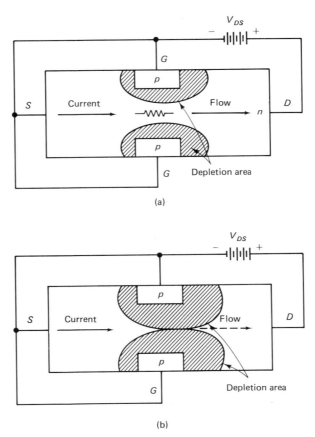

Figure 2.2 Depletion areas of a J-FET caused by the drain-source voltage. (a) Drain-source voltage V_{DS} less than pinch-off voltage V_p. (b) Drain-source voltage V_{DS} greater than pinch-off voltage v_p.

that any further increase in V_{DS} has no further effect on channel conduction. This region we associate with what we call the saturation region.

2.3 OPERATION: ENHANCEMENT MODE

Since enhancement-mode operation is unique to MOSFETs, we shall use the idealized cross section in Fig. 2–3 as our model. In Chap. 1 we associated the insulated metal gate electrode as one plate of a capacitor. When a potential appeared on this plate, an opposite potential charge developed in the substrate directly across the oxide insulation. In our model, a p-channel MOSFET, the n body, naturally provides an excess of electrons that migrate to the oxide–semiconductor interface. If we charge the gate with a positive potential we cause an even greater

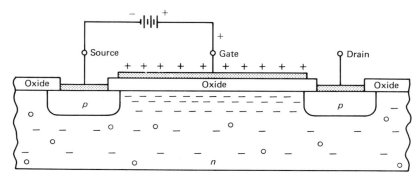

Figure 2.3 P-channel enhancement-mode MOSFET, showing the effect of a positive charge on the gate inducing a heavy *n* layer beneath the oxide.

number of electrons to collect under the gate oxide and no majority-carrier conduction occurs. If, however, we place a negative potential on the gate electrode, the negative charged electrons that previously had collected along the oxide–semiconductor interface will migrate away from this interface.

Once we have increased the negative gate potential to a sufficient level to overcome the excess of electrons characteristic of *n*-type substrate semiconductors, we witness a phenomenon called *inversion*. Inversion simply means that the doped semiconductor material directly under the gate electrode has reversed its polarity. What was once *n*-doped semiconductor by virtue of an impurity implant has become *p*-doped by the excessive depletion of free electrons repelled by the negatively charged gate, as illustrated in Fig. 2-4.

In the example above we should be aware that the opposite holds

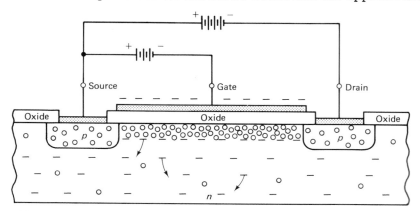

Figure 2.4 P-channel enhancement-mode MOSFET, showing the effect of a negative charge on the gate, which "inverts" the substrate from *n* to *p*.

true for the *n*-channel enhancement-mode MOSFET. Here we would achieve majority-carrier current flow by applying a positive potential to the gate electrode.

2.4 TYPES OF POWER FETs

Despite the apparent plethora of power FETs, often confused by catchy trade names, there remains only two basic types, the depletion-mode J-FET and the enhancement-mode MOSFET. Furthermore, as we will discover in Chap. 4, an optimally designed power FET requires that its channel length be as short as is physically possible, which generally narrows the varieties of FETs to those exhibiting current flow normal to the surface. In other words, we generally find that all power FETs utilize vertical current flow. A common identifier upon which many trade names have been taken for power FETs is that they are called *vertical* FETs.

Not only is our perspective narrowed to vertical construction, but in order for us to achieve high power-handling capability as well as high forward transconductance, low saturation resistance, and superior high-frequency performance, we must arrange for multiple channels in parallel. We are further restricted as to how we may parallel these channels, for it is vital that we optimize what is known as the *figure of merit* for this power FET. What we need to do is to attain the highest possible forward transconductance at both an optimum drain current (g_m/I_D) and with a minimum parasitic capacitance (g_m/C_{in}). Since attaining both simultaneously borders on the paradoxical, achieving an optimized figure of merit is no small task.

The problem of optimizing the figure of merit is basic to FET design. Achieving a high forward transconductance necessitates high drain current. High drain current, in turn, forces the designer to enlarge the geometry (area) and, of course, an enlarged area means that we will be plagued with increased parasitic capacitances. What we find ourselves faced with is what is called the high-frequency problem (Sec. 1.3.1).

Since we are familiar with the work of Teszner and Zuleeg, we shall begin our study of power FETs by focusing our attention on two types of vertical J-FETs, among them Teszner's Gridistor and Zuleeg's MUCH-FET.

Before we focus our attention on the vertical J-FET, we should familiarize ourselves with the term "figure of merit," which appears to be the cornerstone of successful power FET design. The term "figure of merit" originated as an expression to define the gain–bandwidth performance of vacuum-tube amplifiers. There were, in fact, two situations where we would find this expression used. One to compare the performance of steady-state amplifiers; the other to compare transient (fast) amplifiers. In either case the expression was useful in the early days to select the best vacuum tube for the particular design. The utility of this expression has survived the era of the vacuum tube, the bipolar transistor, and now the power FET. We can presume quite correctly that in a steady-state amplifier the product of gain and bandwidth is significant and useful for comparison. What we are saying is simply that our *perfect* amplifier will exhibit constant gain over its full bandwidth. Anything less exhibits a figure of merit less than infinity!

A figure of merit for a transient or fast pulse amplifier is equally significant and useful in evaluating the active device, as well as (as with the steady-state amplifier) the circuit. Transient amplifiers must respond to pulses with fast rise and fall times. Like the steady-state amplifier, we see a critical dependence upon high voltage gain and low parasitic capacitance. In a simple-tuned amplifier, voltage gain is always:

$$\text{gain} = g_m R \tag{2.1}$$

where R is the load resistor and g_m is the transconductance of the active element.

In a transient amplifier we define the rise time of a pulse as

$$\text{rise time} = 2.2RC_{in} \tag{2.2}$$

Combining Eqs. (2.1) and (2.2), we arrive at the expression for the transient amplifier's figure of merit:

$$\text{figure of merit}_{(t)} = \frac{\text{gain}}{\text{rise time}} = \frac{g_m}{2.2C_{in}} \tag{2.3}$$

For the simple steady-state single-tuned amplifier we have Eq. (2.1) to define voltage gain. Bandwidth, considered as the 3-dB rolloff both above and below the center frequency, is;

$$\text{bandwidth} = \frac{1}{2\pi RC_{in}} \tag{2.4}$$

If we combine Eqs. (2.1) and (2.4), we have the alternative expression for a steady-state figure of merit:

$$\text{figure of merit}_{(s)} = \text{gain bandwidth} = \frac{g_m}{2\pi C_{in}} \tag{2.5}$$

Examining Eqs. (2.3) and (2.5), we see that for an optimum figure of merit we need a high value of transconductance and a low parasitic capacitance. This combination is the familiar high-frequency problem that both Teszner and Zuleeg claimed to have solved.

Although a high figure of merit is certainly desirable for all power FETs, we can especially understand its importance in both high-speed switching and in high-frequency applications. When we study Chaps. 9 and 10 we will have an opportunity to compare the performance of those power FETs that exhibit a high figure of merit with other less optimized power FETs.

2.4.1 The Cylindrical Power FET: The Keystone to Success

When we study both Teszner's and Zuleeg's early work, we see a remarkable parallel in their design philosophy simply because both investigators followed closely the work proposed by Wegener, who, in 1959, published a paper entitled "The Cylindrical Field Effect Transistor." In general, Wegener postulated that current saturation in a cylindrical cross section occurs at a lower pinch-off voltage and—and this is of paramount importance—the saturation current is literally orders of magnitude greater than what we had previously experienced from the typical J-FET. His proof of this had been realized from a careful derivation of Poisson's equations. Since it is beyond the scope of this book to delve into the derivations of Poisson's equations, we again invite the reader to research the literature contained in the References concluding this chapter. The elementary cylindrical field-effect transistor that Wegener derived is shown in Fig. 2–5.

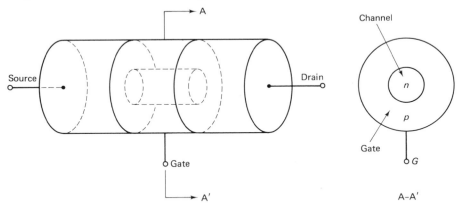

Figure 2.5 Elementary concept of a cylindrical J-FET as proposed by Wegener.

The fundamental advantage that this cylindrical cross-section offers us may be understood as we compare *centripetal striction* with *laminar striction*: the former being that associated with the cylindrical FET, the latter for the planar or laminar FET.

These early researchers reasoned that if FETs were built with equal transverse dimensions and with the same semiconductor material, then the pinch-off voltage for the cylindrical FET would be one-half that of the planar or laminar FET.

Exactly how this cylindrical FET concept aided the early work in power FETs can best be described by comparing the current–voltage characteristics of Zuleeg's MUCH-FET with those of a conventional planar, or laminar J-FET, as shown in Fig. 2-6. If we recognize that transconductance (g_m) is the ratio of a change in drain current to a change in gate input voltage ($\Delta I_D/\Delta V_{GS}$), it is easy for us to see the remarkable improvement in transconductance over that previously attainable when using a conventional planar, or laminar design.

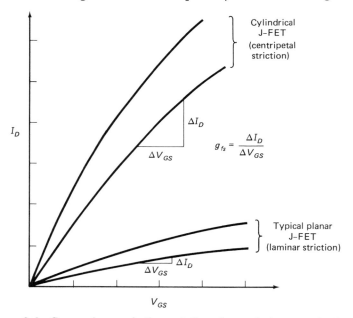

Figure 2.6 Comparison of the relative forward transconductance between the cylindrical and planar J-FET. (© 1965 IEEE. Reprinted with permission from "Ultimate FET Structures," by Esaki and Chang from *Proc. IEEE* (Correspondence), Vol. 53, Dec. 1965.)

Although we have indirectly identified the reason for this much improved transconductance, we should again pause in this review of the power FET to understand the term and, in addition, to appreciate the effects of the *pinch-off voltage.*

2.4.1.1 Pinch-off voltage: what it is and its effect on transconductance. Pinch-off voltage *always* refers to a depletion-mode FET (conversely, an enhancement-mode FET is recognized by its *threshold voltage*, a term that we develop later in this chapter). Were we to apply a positive voltage between the drain and source of the simple n-channel J-FET (which we illustrated in Fig 2-2) and slowly increase this voltage, the drain current would increase in a very predictable manner, as shown in Fig. 2-7.

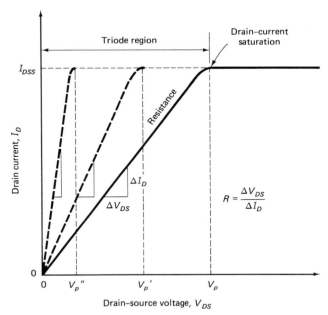

Figure 2.7 Effect of channel resistance R_{DS} as pinch-off voltage is lowered and drain current saturation I_{DSS} remains constant.

As we continue to increase the drain–source voltage (V_{DS}) the depletion region gradually constricts (*laminar striction*), which results in the gradual saturation of the drain current (I_D). The point at which V_{DS} has caused the depletion region to constrict the channel completely, preventing further increase in drain current (drain current saturation), is called the pinch-off voltage, V_p. Beyond that value we can increase V_{DS} but without seeing an appreciable increase in drain current. This effect is clear in Fig. 2–7, where when V_p equals V_{DS}, drain current saturation occurs. This drain current is often referred to as I_{DSS}.

If we examine Fig. 2–7 and, in particular, the triode or unsaturated region, we discover that for a fixed saturation current (I_{DSS}), as the pinch-off voltage drops, the channel resistance also drops. By Wegener's proof that a cylindrical FET offered a pinch-off voltage one-half that of a similarly rated (equal current) planar FET, he also swept away the remaining barrier in the path leading to successful power FET design.

We would need only a rudimentary knowledge of Ohm's law to

recognize that the major hindrance intrinsic to the planar FET was its channel resistance. Channel resistance is the source of several major problems that for decades stymied the successful development of power FETs. For optimum power delivered to a load from a power FET amplifier, we can derive the power equation from Ohm's law,

$$P_{out} = I_D(V_{DS} - V_{DS(on)}) \qquad (2.6)$$

where I_D = drain current

V_{DS} = drain-source supply voltage

$V_{DS(on)}$ = voltage drop across FET

The latter term, $V_{DS(on)}$, results from the effect of the channel resistance.

It is also quite obvious that the current flow through the channel resistance generates heat. The resulting temperature rise leads to a regenerative *thermal* effect if adequate heat sinking is not provided.[1] As the temperature rises, the operating drain current drops (the "fail-safe" or degenerative effect discussed in Sec. 1.2.1.2). With this drop in drain current we have a corresponding drop in forward transconductance. Additionally, as our channel resistance rises, we observe lower gain, not only because of the lowering transconductance caused by heating, but because the channel resistance is a form of degeneration. This degenerative effect can best be understood by the equivalent circuit shown in Fig. 2-8.

Furthermore, we recognize that channel resistance plays a debilitating role in the gain–bandwidth, or figure of merit, of the FET and as

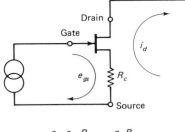

$$A_v = \frac{g_{fs}\, e_{gs}\, R_c}{e_{gs} + i_d R_c} = \frac{g_{fs} R_c}{1 + g_{fs} R_c}$$

Figure 2.8 Schematic diagram identifying the degeneration effect that channel resistance plays in a laminar J-FET.

[1] Heat sinking is a difficult problem and depends to a great extent upon the geometry of the FET itself. Planar construction is generally to be avoided because semiconductor materials have poor thermal properties. Generally, power transistors, both bipolar and FET, are constructed on thinner semiconductor substrates and special precautions are taken to reduce the thermal resistance of the die attach to the package. We address the subject in great detail in Chap. 5.

a consequence limits the high-frequency performance. This is discussed in detail in Chap. 10.

2.4.1.2. Centripetal versus laminar striction.

We have seen from Fig. 2–6 the
benefit of the cylindrical FET as compared to the planar or laminar
FET. We should be able to examine the geometry of a cylindrical FET
and by comparing it to that of the planar FET arrive at a reasonable
understanding of the fundamental advantage offered us by the cylindrical FET.

In Fig. 2–7 we discovered that for a fixed saturation current, as
the pinch-off voltage decreases, the channel resistance also decreases. A
reduction in channel resistance improves the forward transconductance
and, furthermore, becomes a major factor in raising the gain–bandwidth
(or figure of merit) of the FET. We can use Fig. 2–9 to resolve empiri-
cally, to our satisfaction, that the cylindrical cross section does, indeed,
offer a lower pinch-off voltage than does the planar FET of similar

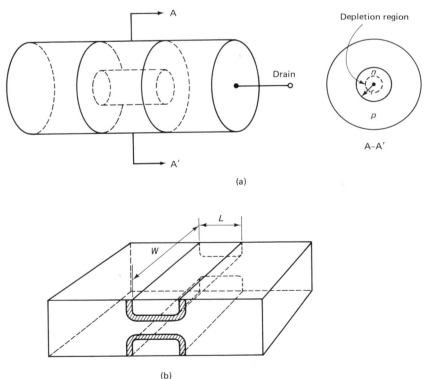

(a)

(b)

Figure 2.9 Simplistic comparison between a cylindrical and planar
geometry J-FET. We see the depletion region of the cylindrical J-FET
(a) constricting more rapidly than within the planar J-FET (b).

cross section and identical saturation current (I_{DSS}).   Since in the
planar FET, W is fixed irrespective of gate bias, as the bias extends the
depletion region under the gate, we find the cylindrical channel constricting at a faster rate than that of the planar FET's channel. Consequently, the cylindrical FET, for the same cross section and tranverse
dimension ($a = r$), exhibits a proportionally higher forward transconductance.

A similar phenomenological analysis can be performed to show an
improved figure of merit based on lower resistance, higher transconductance, and lower gate-source, or input, capacitance.

If we had intuitively expected that the cylindrical FET, constructed in the manner shown in Fig. 2-9 (a), would exhibit poor thermal
properties, we would be correct.   A single-channel cylindrical FET
would be at a severe disadvantage over its planar, or laminar equivalent.
However, a multichannel FET utilizing centripetal striction can overcome this disadvantage if our construction follows the concepts shown
in Fig. 2-10.   In this figure we see the resulting multielement centripetal FET structure as a vertical FET with the drain electrode mounted
directly to an adequate heat sink.

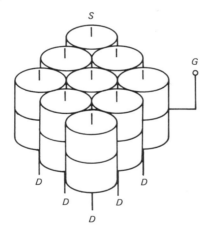

Figure 2.10 By stacking the cylindrical J-FETs on end, we are able
to improve the thermal resistance.

Out of this novel cylindrical geometry concept we are able to
develop a power junction FET (J-FET). We have seen, now for the first
time, a successful solution to the high-frequency problem (see Sec.
1.3.1). The figure of merit that we can now achieve with the cylindrical geometry surpasses by orders of magnitude that which was previously possible with more conventional laminar or planar J-FET. We
saw this improvement in Fig. 2-6, where the transconductance of the
cylindrical J-FET rose far above that of the conventional laminar
J-FET. If we were to assume that the parasitic capacitances of these
two geometries (the cylindrical and the laminar) were equal, the im-

provement in the figure of merit would be obvious. However, being reminded that the cylindrical geometry also offers reduced parasitic capacitances, the figure of merit of the cylindrical J-FET is even more evident.

We will be able to identify several power J-FET designs based upon this novel cylindrical geometry concept as we continue through this chapter. Returning to the early work of Teszner and Zuleeg, we find that the gridistor and the MUCH-FET (*multi-channel FET*) both leaned heavily upon the earlier work of Wegener's cylindrical J-FET.

Since the intent of this book is to develop an appreciation of the performance and application of power FETs, we will limit our study with regard to the constructional details of these and subsequent devices. As we review various power FETs we will examine only those design and constructional details that will help illuminate not only our understanding of their strengths and weaknesses but also their utility in various applications.

2.4.2 The Gridistor

Teszner proposed to combine the advantages of the bipolar transistor and the FET. His goal was simply to improve the figure of merit of the FET to make it more suitable for inclusion into high-speed integrated circuits.

If we examine the simplified cross section of the gridistor, shown in Fig. 2–11, we are able to identify a basic *npn* bipolar transistor structure. Teszner uniquely designed this structure with an embedded grid of *p*-type material, surrounded by *n*-type so that he could effect

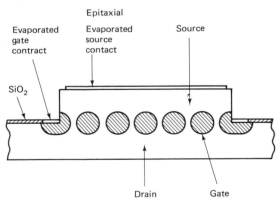

Figure 2.11 Cross section of a "brush" Gridistor structure integrated by epitaxial deposition. [© 1964 IRE (now IEEE). Reprinted, with permission from "Gridistor—A New Field-Effect Device," by Teszner and Gicquel, from *Proc. IRE*, Vol. 52, Dec. 1964.]

either bipolar or FET action merely by the manipulation of the gate bias.

To operate this gridistor as a J-FET we would impress a negative potential on the gate. This, in turn, establishes a depletion region not unlike that which is common to all conventional J-FETs, as shown in Fig. 2-12.

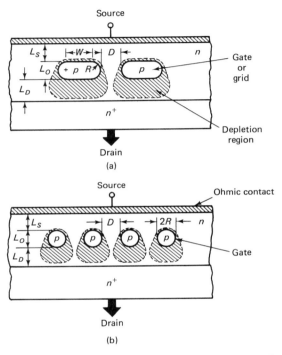

Figure 2.12 Actual (a) and ideal (b) grid structure and depletion region shape near channel pinch-off. (Reprinted with permission from *Journal of Solid State Electronics*, Vol. 10, Zuleeg, "Multi-Channel Field-Effect Transistor," © 1967, Pergamon Press, Ltd.)

Unlike the planar, or laminar J-FET, the gridistor has a very short channel. Where with the planar J-FET we find the channel length to be more or less dependent upon a combination of photolithographic and diffusion techniques, with the gridistor, channel length is limited to the diffusion technique. We can better appreciate the differences between the gridistor and the planar J-FET by comparing Fig. 2-12 with Fig. 1-2(a).

Because of the extremely short channel typical of centripetal striction FETs, the channel of the gridistor does not pinch off as we increase the drain voltage. Consequently, the drain current does not saturate but continues as a function of the drain–source voltage, with the result that the output characteristics resemble those of the triode. Typical of a triode the ratio of incremental drain current to incremental drain–source voltage remains high and the mu, or gain, is low— very much like a power triode.

Teszner's gridistor was a nearly exact analog of a space-charged-limited (SCL) vacuum triode. Because of this analogy, the gridistor exhibits the output characteristics of a triode. Typical output characteristics are shown in Fig. 2–13.

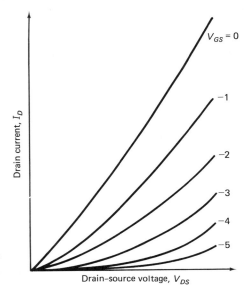

Figure 2.13 Triodelike output characteristics of a gridistor with negative gate voltage.

The principal faults of the gridistor were twofold. For a single channel the maximum realizable forward transconductance was limited to approximately 200 μmhos, this limitation being independent of size. The second failing was its inability to dissipate sufficient power, since, as we have seen, the channel is cylindrical rather than planar. According to Teszner, the power dissipation of a single cell could hardly be expected to exceed 90 to 100 mW.

Early gridistor development offered us remarkable high-frequency performance compared to the planar J-FET. An early data sheet (ca. 1970) announced operation to 2 GHz! This performance is understandable because the gridistor, utilizing the centripetal striction concept, exhibited a reduced transit time, a lower channel resistance, and a low output conductance. These multiple advantages coupled with lower parasitic capacitances, also characteristic of the cylindrical geometry, provided the gridistor with an exceptional figure of merit.

The advantages of the gridistor were not achieved with a single cell. The basic faults had to be overcome before a practical power FET could evolve. From a practical standpoint it was imperative that we increase the power dissipation capability, increase the forward transconductance, and, hopefully, increase the output conductance.

As we mentioned earlier, Zuleeg had proposed and developed prototype power FETs which he called the MUltiple-CHannel FET, or MUCH-FET. A single cell of this MUCH-FET closely resembled the basic concept of the gridistor. Zuleeg's MUCH-FET was, in principle, a practical multicell gridistor (Fig. 2-14) offering high forward transconductance, high current-handling capability (by virtue of many cells in parallel), and high thermal dissipation.

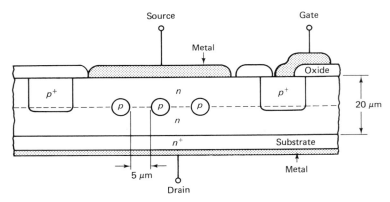

Figure 2.14   Cross-sectional view of the basic MUCH-FET.

The MUCH-FET was designed specifically for high-frequency and high-power operation and early devices (ca. 1966) achieved 2 W at 100 MHz. Zuleeg's projection of power versus the number of parallel channels is shown in Fig. 2-15.

In the intervening years since the development of these transistors we have seen little evidence of their acceptance in industry. Because of their low profile it becomes difficult for us to judge the merits of these two centripetal stricture devices. However, we can fault both devices on at least *one* count. Like all depletion-mode J-FETs, they require two power supplies: one for the drain source voltage, the other for bias. Bias must be of the opposite polarity from that of the drain–source potential to ensure that we can turn the FET OFF.

Nonetheless, we owe a debt to both Teszner and Zuleeg for their

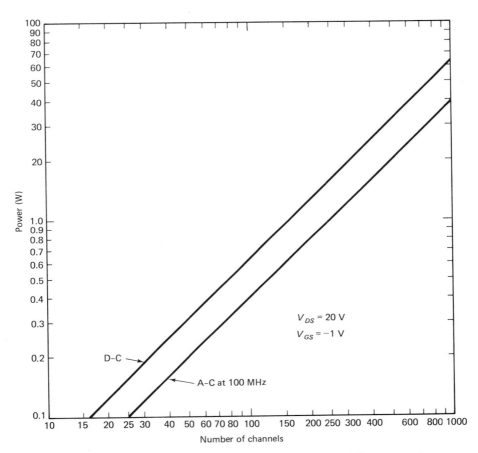

Figure 2.15 Power-handling capability of cylindrical J-FETs versus the number of channels. (Reprinted with permission from *Journal of Solid State Electronics*, Vol. 10, Zuleeg, "Multi-Channel Field-Effect Transistor," © 1967, Pergamon Press, Ltd.)

pioneer work with centripetal stricture FETs. They opened the way for the static induction transistor, to which we now turn our attention.

2.4.4 The Static Induction Transistor

The name *static induction transistor* (SIT) was coined by its inventors to distinguish its mode of operation from that of the *analog transistor* originally proposed by Shockley (1952) and later fabricated as both the gridistor and the MUCH-FET. What principle this name implies is unclear. Many are led to believe that the static induction transistor is nothing more than a highly successful effort by the Japanese in furthering the work of Teszner and Zuleeg.

When we compare the cross-sectional view of the SIT in Fig. 2–16 with either the gridistor (Fig. 2–11) or the MUCH-FET (Fig. 2–14), we

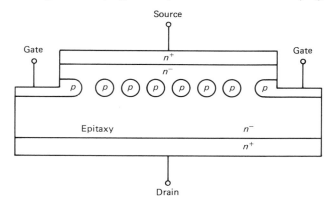

Figure 2.16 Cross-sectional view of the basic static induction transistor (SIT).

must admit that any difference is imperceptible. Although perhaps hotly contested by some, we can identify the SIT as bearing a close analogy with the analog transistor, gridistor, and MUCH-FET.

Nishizawa, the inventor, correctly proposed that drain-current saturation, common among J-FETs, was principally a result of negative feedback stemming from excessive channel resistance. We can visualize his reasoning from Fig. 2–17, where the source resistance is common to both the output and input circuits, thus effecting feedback. As the drain current increases, the gate-to-source feedback voltage

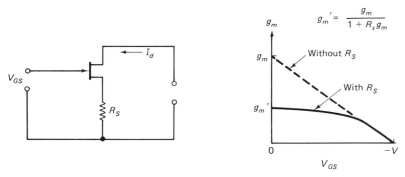

Figure 2.17 Effect of channel resistance (degeneration) on transconductance.

increases proportionally, tending to drive the FET into current saturation.

The SIT, with its very short channel, exhibits a nonsaturating drain current similar to that of the vacuum tube or space-charged-limited triode. We should also take note that both the gridistor and the MUCH-FET had similar output characteristics. We can visualize with greater appreciation the reason for this drastic reduction in channel resistance if we compare the cross section of the SIT in Fig. 2-16 with that of a conventional planar J-FET shown earlier in Fig. 1-2(a).

We must admit to one very unusual feature of the SIT, which also appeared possible with either the gridistor or the MUCH-FET. Both Teszner and Zuleeg identified the phenomenon but apparently did not pursue further study. The SIT, in a general sense, belongs to the family of J-FETs, or depletion-mode FETs (see Fig. 1-1). According to Nishizawa and Ohmi, through the combination of a closely spaced gate structure, clearly evidenced in Fig. 2-16, and a proper impurity concentration within the channel, we can witness the merging, or pinch-off, of the natural depletion region without any bias. Simply stated, the inventors have produced what might be construed as an enhancement-mode J-FET! In other words, we might believe that the SIT could be built exhibiting a normally OFF condition when bias is removed. However, we must move with caution, for if this enchancement-mode SIT is truly in the OFF state, how can we turn it ON? Certainly not by further reduction of the already nonexistent gate voltage as we would expect with an n channel J-FET. To activate this enhancement-mode SIT we must inject minority carriers into the gate. It appears that the normally OFF SIT is tantamount to being a close relative to the npn bipolar transistor. As we might expect, the output characteristics resemble those of the bipolar transistor, as shown in Fig. 2-18.

This "bipolar" SIT has been named the *bipolar-mode static induction transistor* (BSIT). Admittedly, it appears as a very interesting and potentially promising device, but further investigation of this device lies outside the scope of this book.

A major feature of the SIT is its ability to withstand high drain-gate voltages. Ostensibly because of the high velocities imparted to the majority carriers, the drain–drift region can be extended to allow unusually high drain–gate breakdown voltages. Benefits arising from this high standoff voltage include a reduced parasitic drain–gate capacity and a more advantageous design philosophy for power amplifiers. We cover these benefits in detail in later chapters.

The depletion-mode SIT has been fabricated successfully for numerous industrial applications, ranging from high-fidelity audio amplifiers to high-frequency power amplifiers. Because of the high break-

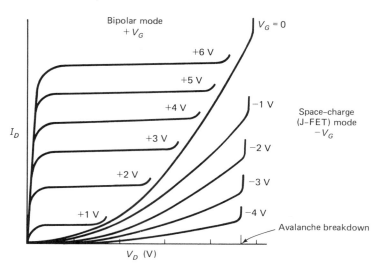

Figure 2.18 Comparison of the output characteristics of a static induction transistor with negative and positive gate bias. (R. Zuleeg, U.S. Patent 3,409,812, "Space-charge-limited current triode device," Nov. 5, 1968.)

down voltages available with this technology, we shall, in all probability, see the SIT as a strong contender in many power designs. To be sure there will be strong competition from other power FETs, which we describe in detail later in this chapter.

2.4.5 The Insulated-Gate Power FET

We need not look further to find exciting advances in power FET technology. Although the power J-FET preceded the insulated-gate technology by nearly two decades, it is to the latter that our attention must be focused.

Common to all the J-FETs that we have studied heretofore, each was controlled by an element called a *gate*. This gate was, for all cases, a reverse-biased *p-n* junction. Majority-carrier current flow was controlled solely by the action of a depletion field. This depletion field resulted from the combined interaction of the *p-n* junction and the applied gate bias voltage.

The insulated-gate FET operates on entirely different principles. We find the *p-n* junction (gate) replaced by a metal plate separated from the semiconductor by a thin insulating film. Low-frequency[2] power FETs have relied upon the advanced status of silicon tech-

---

[2]Frequencies generally below 1.5 GHz. For higher frequencies the semiconductor has favored gallium arsenide and its various compounds.

nology, so we can expect this insulating film to be silicon dioxide. The insulated-gate FET (or IGFET) is widely known as a *metal-oxide semiconductor* FET or MOSFET.

As we search the field of power MOSFETs, we find that irrespective of their construction (which is described in detail in Chap. 3), all of them are what are commonly known as *enhancement-mode* MOSFETs. Enhancement mode means that only by the application of the proper bias to the gate electrode can we induce, or "enhance", the electron (current) flow through the channel. Such a power MOSFET is considered to be OFF (nonconducting) when the bias is removed from the gate.

Since we reviewed the basic MOSFET principles in Secs. 1.1 and 2.3, for the moment we focus our attention on the mechanism of *inversion* and its associated *threshold voltage*.

In our illustration of a *p*-channel MOSFET in Sec. 2.3, we visualized, through the help of Fig. 2–3, the basic operating principles of the enhancement-mode MOSFET. Power MOSFETs are generally of two types: *n*-channel and *p*-channel. Although our illustration was descriptive of a *p*-channel MOSFET, the *n*-channel may be easily understood if we simply invert the semiconductor dopants and reverse the gate polarity from negative to positive. Again, as before, we have an enhancement-mode, normally OFF, MOSFET. For either the *p*-channel or *n*-channel a gate bias of the proper polarity must be impressed to turn the MOSFET ON. For the *p*-channel, this gate potential must be *negative*; for the *n*-channel, *positive*.

2.4.5.1 Inversion. If we apply a voltage across this *n*-channel MOSFET—from drain to source—no current will flow. We have, in effect, a pair of series-connected back-to-back *np–pn* diodes, as shown in Fig. 2–19. To induce current flow we must impress a positive voltage onto the gate electrode. Immediately, an electric field is set up

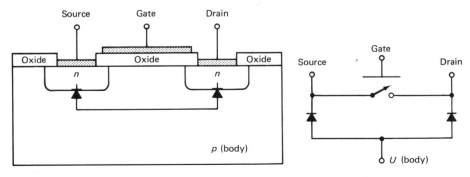

Figure 2.19 Equivalent circuit of the OFF state of an enhancement-mode MOSFET.

across the dielectric, attracting free electrons resident within the *p*-doped semiconductor substrate (or *body*).   As these free electrons collect along the surface immediately beneath the positively charged gate electrode, that portion of the substrate *inverts*.   That is, once sufficient free electrons have gathered, the once *p*-doped (acceptor) channel becomes an electron-enhanced *n*-channel.   Current is now free to flow between the source and drain and the MOSFET is ON.

2.4.5.2 Threshold Voltage.   It is important that we recognize that turning the MOSFET ON is a gradual process.   A typical transfer characteristic for an *n*-channel enhancement-mode MOSFET is shown in Fig. 2–20, where we can see a smooth transition from nonconducting to conducting.   It is quite obvious from this illustration that an immediate application of a positive gate voltage does *not* induce immediate conduction.   This finite gate voltage that appears to be needed is called the *threshold voltage* (V_{th}).

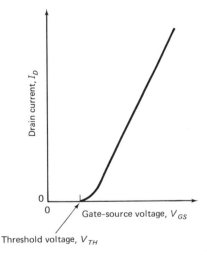

Figure 2.20   Typical static transfer characteristic of an enhancement-mode *n*-channel MOSFET.

The amount of gate bias needed to begin conduction—the threshold voltage—depends upon several factors: the concentration of impurities (dopant) in the semiconductor substrate (body), the magnitude of charge resident in the gate-insulating oxide, and the type of metal used for the gate electrode itself.

Occasionally, with low-breakdown-voltage MOSFETs such as we might find in integrated circuits, the threshold voltage may show a dependency upon the applied drain–source voltage.   This dependency results from interactions of the drain depletion region extending into the *p*-doped channel, which alters the substrate impurity concentration.   We find that the high-voltage, high-power MOSFETs are free of this problem because of their wide *drain–drift* region inherent in their design.

Aside from the several internal factors, threshold voltage is critically dependent upon operating temperature. As we raise the temperature of the MOSFET semiconductor, the threshold voltage falls (a coefficient of approximately -5mV per °C). We must exert great care to prevent the threshold voltage from falling *below* zero; otherwise, our normally OFF enhancement-mode MOSFET will become abnormally ON.

Careful design of the MOSFET will allow us to set the threshold voltage so as to make the enhancement-mode, normally OFF, MOSFET compatible to many forms of logic circuitry. These applications are explored in depth in Chap. 8.

2.4.6 The Short-Channel MOSFET

Before the excitement of power FETs, the common MOSFET was the long-channel planar variety shown in Fig. 2-3. It was typical to fabricate this type of MOSFET using photolithographic techniques, which often limited channel lengths to at least 5 μm. In recent years we have seen increasing emphasis on using MOSFETs in integrated circuits. Because of increasing complexity and accompanying densities we have been forced to consider designing MOSFETs with finer line widths and shallower diffusions. Partly as a fallout of this advancing art, we have seen the successful fabrication of the short-channel MOSFET.

Earlier experiments with MOSFETs quickly identified several problems. Planar construction using photolithographic masking required massive geometries to handle the necessary currents for power applications. Coupled with this, the parasitic capacitances rose out of proportion, reducing the MOSFET's effective gain–bandwidth and, consequently, their speed. These large geometries suffered from high channel resistance and excessive losses ($I \times R$) which were compounded by poor thermal dissipation. The abnormally high channel resistance inhibited both transconductance and gain. Physical problems plagued the manufacturer of these large geometries, not the least being the exorbitant costs resulting from poor yields and extreme difficulty in die attach, which further compounded their reliability.

From this list of seemingly insurmountable problems it takes little imagination for us to acknowledge that for all the innovations and modifications we might try, the planar, photolithographic MOSFET simply will not work as a viable power transistor. Out of this, the short-channel vertical MOSFET evolved.

Photolithographic masking technology has probably taken us as

far as we can go in reducing channel lengths. A novel concept, not requiring close-tolerance masks, is a technology identified as *double-
diffused MOS* or DMOS. Another equally novel technology is the *V-
groove MOS* or VMOS. The former we can identify as basically a planar
process, the latter can only be implemented as a vertical process.

2.4.6.1 Double-diffused MOSFETs.

Perhaps we should first address
the question of why a short channel is required. Earlier in this chapter
we recognized the importance of a high figure of merit that necessitated
high forward transconductance and low parasitic capacitance. Since the
channel length of a MOSFET is inversely proportional to transconductance and proportional to ON resistance, we can immediately recognize
the importance of a short channel. Furthermore, we need little imagination to envision that a small geometry that results from a short
channel will also have lower parasitic capacitances. Consequently, it is
to our distinct advantage to focus our attention on the successful fabrication of short-channel MOSFETs.

The double-diffused MOSFET was one of the earliest successful
efforts in short-channel MOSFET technology. DMOS has undergone
continual refinement and today we find it standing as one of the major
power FET technologies.

The basic DMOS structure can easily be visualized and compared
with the conventional planar MOSFET with the help of Fig. 2-21.
Here we see an n-channel MOSFET with the desired short-channel and,
in addition, an n^--doped drain-drift region extending to the highly
doped n^+ drain. The name "DMOS" was coined from the sequential
manner in which the p-doped (body) diffusion was followed by a highly
doped n^+ source diffusion. This short channel results through careful
control and placement of the second (n^+) diffusion. In Chap. 3 we
explore in greater depth the details and options available to us in the
manufacture of DMOS power transistors.

As a result of the short channel, we find some interesting electrical
characteristics, the most noteworthy being a phenomenon called
electron (or *carrier*) *velocity saturation*. Carrier saturation is evident
in at least two ways. First, as shown in Fig. 2-22(a), we have what
appears as a linear static transfer characteristic (nonlinear before velocity saturation occurs). Second, we can observe from Fig. 2-22(b) a
saturation of the transconductance where regardless of increases in
either gate bias or drain–source voltage, no further change in transconductance is evident.

Were we interested in using DMOS in high-frequency applications,
the transconductance would need to be optimized [see Eq. (2.5)]; and
since transconductance is proportional to carrier speed, we can expect

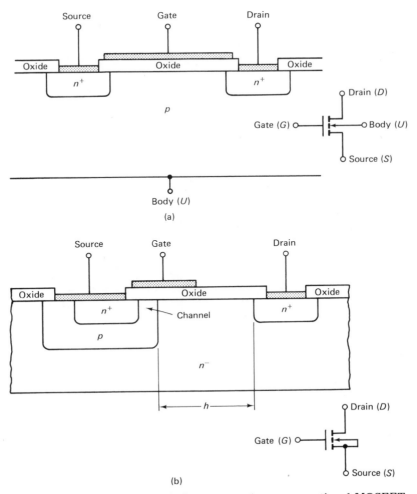

Figure 2.21 Cross-sectional views comparing a conventional MOSFET (a) and a short-channel double-diffused planar MOSFET (b).

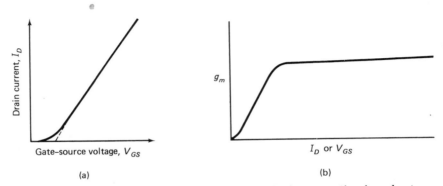

Figure 2.22 Saturation effect caused by velocity saturation in a short-channel MOS. (a) Transfer characteristics. (b) Transconductance.

correctly that the maximum frequency f_t is inversely proportional to channel length:

$$f_t = \frac{v_s}{2\pi l} \tag{2.7}$$

where v_s = carrier velocity, cm/s
      l = length of the channel, μm

　　Having touched briefly on the reasons for the short channel, we need to understand the purpose for the drain-drift region. The design of power MOSFETs involves compromise. For optimum transconductance and low channel resistance we need a short channel which, as we noted previously, also contributes to lower capacity and consequently an improved figure of merit. Yet, if only by intuitive reasoning, we can deduce that a short channel between source and drain will have a gross effect upon the voltage breakdown characteristics. Such reasoning makes us conscious of a trade-off between transconductance, channel resistance, and drain current and the breakdown voltage. As the drain–source voltage rises, the drain–drift region of a low n-doped epitaxy allows the drain–channel depletion layer to spread mainly in this drift region, as illustrated in Fig. 2–23. The penetration of this depletion

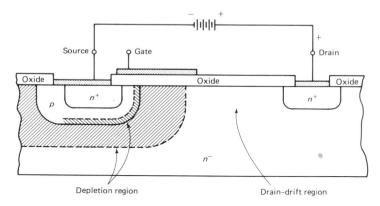

Figure 2.23 The depletion region in the drift region of a double-diffused MOSFET.

layer is largely a function of the ratio of doping concentrations across the reverse-biased p-n junction. The more lightly doped the region, the greater the penetration. In this DMOS structure, our drain–drift region is lightly doped (n^-) and as a consequence a much larger portion of the depletion layer is formed there. It is because of this lightly doped drift region that we find the threshold voltage independent of the drain-source voltage (see Sec. 2.4.5.2).

As we increase the drain–drift region (h in Fig. 2-21(b)), we raise the breakdown voltage but at the expense of added drain–source resistance. This drift–region resistance is seen to increase as

$$r_d \propto BV^{2.5 \text{ to } 2.7} \tag{2.8}$$

For us to reduce the drain–source resistance of a high-voltage break-down short-channel MOSFET structure, we would have to increase the number of parallel channels, thus adding to the chip size and cost.

A high-breakdown-voltage capability does not develop simply as we increase the drain–drift region. Although beyond the scope of this book, we should note that to preserve the high voltage capability, the depletion fields within the drain–drift region must be uniform and show broad-radii curves, as illustrated in Fig. 2-24. Several solutions have been advanced to accomplish this, such as implanting guard rings and adding field plates. In the planar DMOS shown in Fig. 2-21, the gate metal itself acts as a partial field plate to help preserve the high-voltage capability.

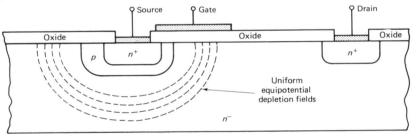

Figure 2.24   By establishing a uniform depletion field we are assured of good voltage breakdown characteristics.

The maximum current capacity of DMOS is dependent upon the root ratio of maximum thermal dissipation to total resistance (if we exclude the effects of current through the metalizations and lead bonds). As we increase our breakdown voltage we also increase the drift resistance [Eq. (2.8)], but drift resistance contributes to the thermal problem differently than does channel resistance. Once the drain current passes through the channel, we see it fanning out into the drift region, and this increased area aids in its dissipation.

An enlarged drain-drift region also increases transit time—the time for majority charges to reach the drain. This added delay can affect our performance both in high-speed switching and in high-frequency applications.

2.4.6.2. The parasitic power bipolar transistor.   We must not be lulled into thinking that the short-channel structure is, of itself, ideal. Conceptually, perhaps; practically, not quite! Every short-channel MOSFET is plagued with a parasitic bipolar transistor in shunt with the MOSFET. If we carefully study Fig. 2-21(b), comparing it with the conventional planar MOSFET [Fig. 2-21(a)], we can at once see the problem. To preserve the features and benefits of the power MOSFET it is absolutely essential that we mute any overt (and covert) activity arising from this parasitic bipolar transistor. For an n-channel short-channel MOSFET this shunt parasitic is an npn bipolar transistor. Were we to inspect the metal mask of virtually *any* power MOSFET we would see an ohmic metal contact bridging between the source and the p-doped body (Fig. 2-25). Because of this bridge we have simply shorted the parasitic bipolar transistor's base to its emitter. *Theoretically* at least, we have effectively muted the parasitic transistor.

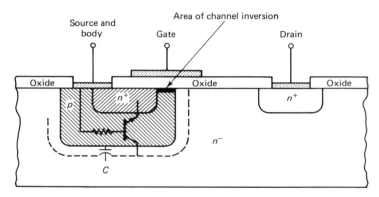

Figure 2.25 By bridging source (n^+) to body (p) we are able to mute the parasitic npn bipolar transistor.

When an ohmic short exists on a planar device such as the MOSFET shown in Fig. 2-25, we may be fooling ourselves if we think we have muted the parasitic bipolar transistor. We have not previously discussed the resistivities of the different diffusions; it was not our intent to do so. However, we do acknowledge the bulk resistivity of silicon, so we are able to visualize that the various doped impurities, whether n or p, also have finite resistivities. It is common to refer to these resistivities as having so many *ohms per square*. A closer examination of Fig. 2-25 reveals that the bridged short between the p body and the n^+ source lies some distance from the area of channel inversion. The danger we face can be catastrophic destruction! This finite resistivity appears as a series resistance between the base and emitter of our parasitic bipolar transistor. If we can, in some way, cause an electric charge to bridge this resistor, we might activate the parasitic bipolar transistor. There are several ways in which we might induce this charge. Taking one way, from Fig. 2-25 we can recognize that a parasitic capacitor exists between the p-base and the n^- drain–drift region. Under proper conditions this capacitor can transfer a charge directly to the "base" of our parasitic bipolar transistor. A fast rise or fall in drain voltage (slew rate) caused, for example, in switching a high current might be enough to trigger the parasitic transistor ON. What is of paramount concern is a phenomenon that manufacturers of power FETs prefer not to discuss: *secondary breakdown*. We find, however, as we research the voluminous literature on this subject that there are conflicting opinions as to the specific cause of this destructive phenomenon. Readers are encouraged to check through the References at the end of the chapter and pick a theory that best suits their philosophy!

In any event, we try to prevent activating this parasitic bipolar transistor. But if we do succeed, the most visible result is an effect called secondary breakdown, which more often that not results in catastrophic destruction of the MOSFET.

2.4.6.3. The parasitic source-drain diode. Not all parasitic elements are necessarily undesirable. In the short-channel MOSFET we have the p (base) n^- (drain-drift) junction, which both looks and acts like a reverse-biased diode across the drain to source. A schematic diagram showing both the parasitic bipolar transistor and this parasitic diode is shown in Fig. 2-26. Where the parasitic bipolar transistor is definitely undesirable, the diode is, for the most part, highly beneficial.

By careful design of the MOSFET the recovery time of this diode can be made sufficiently swift so that it can perform a very useful function in protecting the MOSFET when used to switch inductive loads. More will be said about this in Chaps. 4 through 6.

An undesirable effect of this diode becomes evident if we try using the MOSFET as a series-connected analog switch. The ON condition is no problem, but the OFF state only results in half-wave rectification of the incoming analog signal, as illustrated in Fig. 2-27. We cover this problem and its unique solution in Chap. 9. Returning to our discussion on DMOS we discover that as we opt for higher and higher breakdown voltages, the drain–drift region soon becomes prohibitively large, resulting in chips that simply cannot offer cost-effective yields. One solution that has found wide acceptance is the vertical DMOS structure.

2.4.6.4 The vertical DMOSFET. Where the lateral high-voltage DMOS became unduly large, the vertical structure restored the chip size to where our costs could match those of bipolar transistors of similar ratings. Today, all high-voltage, high-power DMOSFETs are constructed using the vertical concept with the source and gate situated on the top side and the drain on the underside of the chip as shown in Fig. 2-28.

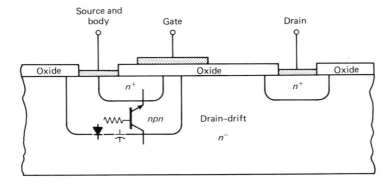

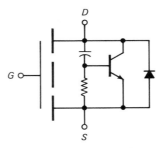

Figure 2.26 Cross-sectional view showing the parasitic *npn* bipolar transistor and drain-source diode. The diode is, in reality, the base–collector diode of the parasitic bipolar transistor, which is shown separately for clarity.

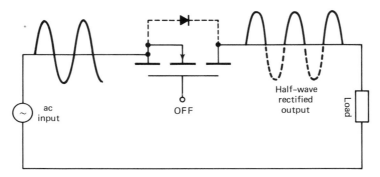

Figure 2.27 Effect of the source–drain diode when using the MOSFET as a series switch. The diode acts as a half-wave rectifier to the analog input signal, resulting in the waveform shown.

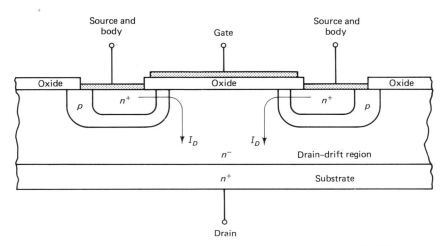

Figure 2.28 Cross-sectional view of the basic vertical double-diffused MOSFET.

Using the vertical structure with the added feature of some type of field limiting, we find the vertical DMOS (VDMOS) capable of withstanding extreme voltages. Some devices are crowding the kilovolt range.

Operationally, there is little to differentiate this vertical structure from its planar or lateral counterpart. But there are some advantages other than breakdown voltage that we need to note. A power transistor that can boast of no secondary breakdown will capture everyone's attention. Vertical DMOS, if properly designed and constructed, can make that boast!

Another advantage of vertical DMOS construction is, as we noted earlier, its cost effectiveness in preserving chip size. The ultimate acceptance of any device lies in its cost–performance trade-offs. We can understand that large geometries have lower yields. If we are to substitute power FETs for the ubiquitous power bipolar transistor effectively, we must show favorable cost-performance. The vertical DMOS structure surely does this.

2.4.6.5 The V-groove MOSFET (VMOS). Another short-channel power FET is the *V-groove* MOSFET, which, as we see in Fig. 2–29, can only be fashioned as a vertical structure. As we study this figure we might note how the basic structure compares to a double-diffused epitaxial bipolar transistor. For an n-channel MOSFET, both begin with an n^+ substrate and an n^- epitaxial into which we first diffuse a p and then an n^+ layer. All similarity ends with the diffusion of this n^+ layer. The V-groove extends through and bisects both the n^+ and p diffusions, terminating in the n^- epitaxial layer. After oxide and metal

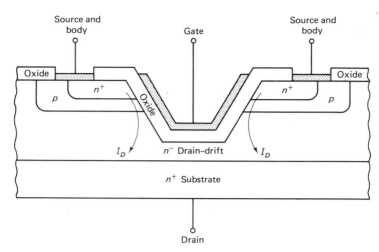

Figure 2.29 Cross-sectional view of a metal gate/vertical V-groove MOSFET.

overlays, we find an insulated-gate power MOSFET that closely resembles the DMOSFET that we studied earlier.

Operating exactly as the DMOS, a positive gate potential inverts the *p* channel, resulting in an uninterrupted, low-resistance current flow between source and drain. Of special interest we should be quick to see that *two* current paths have been established for each V-groove. It is this dual current path that makes VMOS one of the most cost-effective power FETs available.

So far in our study of both the vertical DMOS and the V-groove MOSFET, we have focused our attention on *how they work* rather than on how they are made. The latter we explore in Chap. 3. However, we should note that the metal gate is, for the most part, passé, having been replaced by silicon gate technology. There are a few isolated strongholds where the metal gate dominates, in particular for high-frequency applications, where the series resistance inherent in silicon gate technology would be undesirable.

2.5 WHICH POWER FET DO I CHOOSE?

As we wrap up this chapter, we should take a look at all the power FETs and try to assess where each type would benefit.

We did not discuss complementary power FETs, that is, FETs of opposite polarity. Operationally, they perform in like fashion, except with all voltages reversed. A *p*-channel FET would have a negative drain voltage. An enhancement-mode, *p*-channel MOSFET would turn ON with a negative voltage applied to its gate.

There is an expanding market for p-channel, enhancement-mode, short-channel MOSFETs. Recognizing that the mobility of p-doped silicon is considerably less than that of n-doped silicon, we must expect that a true complement is impossible. Generally, the match is for equal ON resistance and transconductance. Capacity unbalance is the penalty. Consequently, in high-current CMOS applications, one will need heavier drive requirements for the p-channel than for its n-channel complement.

As we view the expanding world of power FETs we see increasing interest in the *static induction transistor* (SIT) in high-frequency applications. Nishizawa has claimed that the SIT will prove useful as a high-power, high-frequency source, and some speculate replacing the magnetron in the microwave oven. The unsaturated drain-current characteristic of the SIT suggests a potentially useful application in linear, or Class A amplifiers.

High-voltage SITs are commercially available for such diverse applications as audio-amplifier output stages and motor controllers. Microwave versions have reached the market boasting of output powers in excess of 100 W at 1 GHz.

The power MOSFET will undoubtedly find wider and more popular use. As we witness the advances in power MOSFET technology, we can only guess when and where the plateau will be reached: that is, where the MOSFET will find its maximum voltage and current and power.

If we were to compare the V-groove MOSFET (VMOS) with the vertical DMOS (VDMOS) we might be hard-pressed to answer our question. But certainly the answer will be influenced by our application.

Many researchers have ventured to state that VMOS will offer the lowest ON resistance, whereas VDMOS will provide the highest stand-off voltage. The real crux, however, will be the *safe operating area* (SOA)—which MOSFET will provide fail-safe performance at the highest voltage and greatest current. Simultaneously, the V_{SAT} or $R_{DS(ON)}$ remains an important parameter that for many applications will always tilt the decision.

In high-frequency applications we find the mobility of silicon to be the limiting factor (hence the popularity of such exotic compounds as gallium arsenide and indium phosphide for microwave applications) and because of the crystalographic orientation of the DMOS process—offering a 25% improvement over VMOS—DMOS may prove superior at UHF. Nonetheless, at HF (2 to 30 MHz) through VHF (30 to 300 MHz) we may find VMOS dominant in the market simply because of its exceptionally low V_{SAT} (actually $R_{DS(on)} \times I_D$) at moderate to low drain voltages.

As the reader finishes this chapter hindsight will have offered us

20-20 vision, and the contest between VDMOS and VMOS may have already been won.

REFERENCES

DECLERCQ, M., and J. PLUMMER, "Avalanche Breakdown in High-Voltage DMOS Devices," *IEEE Trans. Electron Devices*, 23 (1976), 1-4.

GHANDHI, S. K., *Semiconductor Power Devices*, New York: Wiley-Interscience, 1977.

HU, C., "A Parametric Study of Power MOSFETs," *Conf. Record*, Power Electronics Specialists Conference, San Diego, Calif., 1979.

KRISHNA, SURINDER, "Second Breakdown in High Voltage MOS Transistors," *Solid-State Electronics*, 20 (1977), 875-78.

LEHOVEC, K., and R. MILLER, "Field Distribution in Junction Field-Effect Transistors at Large Drain Voltage," *IEEE Trans. Electron Devices*, 22 (1975), 273-81.

LIDLOW, A., T. HERMAN, and H. COLLINS, "Power MOSFET Technology," *Technical Digest*, IEEE Electron Devices International Meeting, Washington, D.C., 1979, pp. 79-83.

LISIAK, K., and J. BURGER, "Optimization of Nonplanar Power MOS Transistors," *IEEE Trans. Electron Devices*, 25 (1978), 1229-34.

MOCHIDA, Y., et al., "Characteristics of Static Induction Transistors: Effects of Series Resistance," *IEEE Trans. Electron Devices*, 25 (1978), 761-67.

NISHIZAWA, J., et al., "Bipolar Mode Static Induction Transistor," *Digest of Technical Papers*, 11th Conference on Solid State Devices, Tokyo, Japan, 1979, pp. 189-90.

NISHIZAWA, J.,T. TERASAKI, and J. SHIBATA, "Field Effect Transistor versus Analog Transistor (Static Induction Transistor)," *IEEE Trans. Electron Devices*, 22 (1975), 185-197.

NISHIZAWA, J., and K. YAMAMOTO, "High-Frequency High-Power Static Induction Transistor," *IEEE Trans. Electron Devices*, 25 (1978), 314-22.

OHMI, TADAHIRO, "Power Static Induction Transistor Technology," *Technical Digest*, IEEE Electron Devices International Meeting, Washington, D.C., 1979, pp. 84-87.

OHMI, TADAHIRO, "Punching through Devices and Its Integration," *Technical Report 43*, Research Institute of Electrical Communications, Tohoku University, Sendai, Japan, 1979.

RITTNER, E., and G. NEUMARK, "Theory of the Surface Gate Dielectric Triode," *Solid-State Electronics*, 9 (1966), 885-98.

SALAMA, C., and J. OAKES, "Nonplanar Power Field-Effect Transistors," *IEEE Trans. Electron Devices*, 25 (1978), 1222-28.

Special Issue on Power MOS Devices, *IEEE Trans. Electron Devices*, 27 (1980), 321-400.

TEMPLE, V., and P. GRAY,"Theoretical Comparison of DMOS and VMOS Structures for Voltage and ON Resistance," *Technical Digest*, IEEE Electron Devices International Meeting, Washington, D.C., 1979, pp. 88-92.

TESZNER, S., "High Frequency and Power Field Effect Transistor with Mesh-like Gate Structure," U.S. Patent 3,274,461 (Sept. 20, 1966).

TESZNER, S., "High Power Field-Effect Transistor," U.S. Patent 2,930,950 (Mar. 29, 1960).

TESZNER, S., and R. GICQUEL, "Gridistor—A New Field-Effect Device," *Proc. IEEE*, 52 (1964), 1502-13.

WEGENER, H. A. R., "The Cylindrical Field-Effect Transistor," *IEEE Trans. Electron Devices*, 6 (1959), 442-449.

"Yamaha et les transistors de puissance à effect de champ," *Revue du SON*, 256/257 (Aug-Sept. 1974), 74-77.

ZULEEG, R., "High-Frequency Field-Effect Triode Device," U.S. Patent 3,381,187 (Apr. 30, 1968).

ZULEEG, R., "Multi-Channel Field-Effect Transistor Theory and Experiment," *Solid-State Electronics*, 10 (1967), 559-76.

ZULEEG, R., "Multichannel Junction Field-Effect Transistor and Process," U.S. Patent 3,967,305 (June 29, 1976).

ZULEEG, R., "Planar Multi-Channel Field-Effect Triode," U.S. Patent 3,381,188 (Apr. 30, 1968).

ZULEEG, R., "A Silicon Space-Charge-Limited Triode and Analog Transistor," *Solid-State Electronics*, 10 (1967), 449-60.

ZULEEG, R., "Space-Charge-Limited Current Triode Device," U.S. Patent 3,409,812 (Nov. 5, 1968).

ZULEEG, R., "A Thin-Film Space-Charge-Limited Triode," *Proc. IEEE*, 54 (1966), 1197-98.

ZULEEG, R., et al., "Electrical Characteristics of the Multi-Channel Field-Effect Transistor," *NEREM Record*, Chicago (1966), pp. 156-57.

three

Fabrication of Power FETs

3.1 INTRODUCTION

In Chap. 2 we studied several types of power FETs, concentrating only on those that have successfully captured the attention of industry. There are many variations we chose to ignore, important as they may be, simply because they have not become attractive to the industry. Furthermore, the purpose of this book is the application of power FETs, and naturally we can address only those that are commercially available. Although the gridistor and the MUCH-FET never caught the attention of industry, they are nonetheless important predecessors to the static induction transistor, so we included them in Chap. 2. Those that we did ignore as well as those that we selected are all basically either J-FETs or insulated-gate MOSFETs. Our family of FETs remains as we saw in Fig. 1–1.

There are many ways in which we could build a power FET. As we read through this chapter we must bear in mind that it is not our intention to be exhaustive or, for that matter, even entirely descriptive or accurate. Our goal is simply to compare the fabrication of some of those power FETs that we studied in Chap. 2. It would be naive of us to believe that fabrication processes are as simple as we have shown them to be in this chapter. These are oversimplified and not practical in today's world.

There is within the J-FET family the static induction transistor (SIT), which appears to be gaining in popularity, especially overseas. Among those that are within the family of insulated-gate FETs we have three basic types: the vertical double-diffused MOSFET (VDMOS), the V-groove MOSFET (VMOS), and the meshed-gate MOSFET. We did not touch on the latter in Chap. 2 as its characteristics closely resemble those of both the VDMOS and the VMOS.

We should be especially aware that what we are about to study is at best a very superficial glimpse into the fabrication of power FETs. Many critical engineering decisions precede even the most basic design long before fabrication starts. One such decision of paramount importance is the manufacturing costs. These costs can be directly related, of course, to the complexity of fabrication; the number of masks, the type and method of doping or implanting, the passivation treatment, and although certainly not the last, the type of metal deposition laid over the surface of the chip for electrical contact to the source and gate (the drain being on the bottom of the vertically constructed power FET). Critical to our overall costs is the physical size (surface area) of the finished semiconductor chip with respect to the size of the wafer. In Fig.

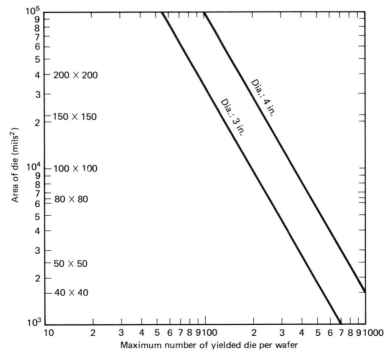

Figure 3.1 Plot showing the maximum number of *yielded* die per wafer size. This clearly shows the advantage of beginning with large-diameter wafers.

3-1 we are afforded some insight of typical yields based upon die and wafer size. This graph, however, does not consider *electrical* yields. The basic design of the power FET, resulting in the most straightforward fabrication steps, aids in establishing electrical yields. Were we to cost out the fab expenses we would certainly have a greater appreciation of chip yields and their ultimate impact both on manufacturing costs and the eventual sale price.

Although it is quite outside the scope of this book to review the physics of semiconductor technology that must be understood to achieve optimum performance, we will, nonetheless, try to point out some of the more apparent problems facing the designer as we proceed through these fabrication steps.

3.2 THE STATIC INDUCTION TRANSISTOR

In Chap. 2 we learned that the static induction transistor has a wide range of potential applications. Not only do we see it used in the power output stage of an audio amplifier but we will find it more and more in high-frequency applications as the technology becomes more refined.

An easy trap that we must be sure to avoid is to mistakenly believe that *a* static induction transistor can work equally well anywhere within the spectrum. As we examine in some detail in this chapter, the SIT can be fabricated in a number of ways to offer optimum performance either in industrial applications, or as a power device in an audio amplifier or in high-power, high-frequency power amplifiers.

For industrial applications we can presume that the SIT should provide a high standoff voltage capability, perhaps several hundred volts, and be capable of controlling many amperes of current. On the other hand, for use in the power stage of a high-frequency amplifier, the standoff voltage requirement may be of little consequence, but the need for low interelectrode parasitic capacitance might be critical.

We begin our trek through the fabrication processes of an industrial version of the SIT using the series of illustrations provided in Fig. 3-2.

Common to all the power FETs that we cover in this book, the SIT fabrication begins with an n-doped silicon wafer substrate [Fig. 3-2(a)]. A typical silicon wafer might be 76 to 100 mm in diameter and range from 0.4 to 0.9 mm thick. This substrate would be heavily doped n^+; for example, we might dope it with 5×10^{19} atoms of arsenic per cubic centimeter (hereafter we identify the impurity concentration as atoms/cm^3).

The breakdown voltage of any of the power FETs that we will examine in this chapter depends to a great extent upon both the doping

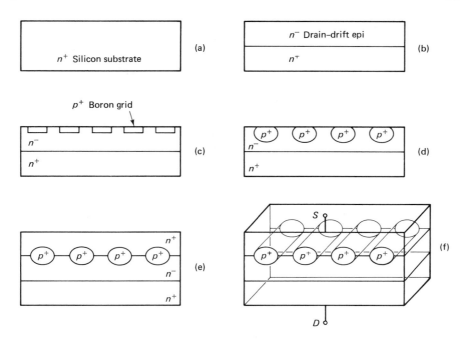

Figure 3.2 Basic construction steps involved in fabrication of a static induction transistor.

concentration as well as the thickness of the epitaxial (epi) layer [Fig. 3-2(b)]. This epi is grown on the topside surface of the silicon wafer substrate. Continuing our example, the doping of this n^- epi would range around 1 to 5×10^{13} atoms/cm^3.

Our next step is to place the gate structure. By suitable masking, oxidation, and etching steps we are able to lay a p^+ boron-doped grid-like structure over the epi, as shown in Fig. 3-2(c). As the wafer passes through the diffusion furnace, the boron diffuses deeply into the n^- epi, giving a profile similar to Fig. 3-2(d). As we continue in our example, the boron concentration would average 1 to 3×10^{19} atoms/cm.^{3}

Following the placement of this p^+ gate structure we now grow still another n epi, which, during the process, causes the p^+ gates to diffuse laterally, giving rise to the now modified view shown in Fig. 3-2(e). With a very shallow second epi layer we have all but concluded the fabrication of the SIT. This final epi, in our example, might have an arsenic concentration of 1 to 3×10^{15} atoms/cm^3.

We now come to the final step involving the deposition of metal for source and gate ohmic contact on the top surface and drain metal on the backside [Fig. 3-2(f)]. Some manufacturers use aluminum for all top surface metal, whereas manufacturers of high-frequency power transistors generally use refractory metals such as gold. For small-area transistor chips the underside drain contact would, in all probability, be gold, which for die attach to most headers can be easily accomplished using gold preforms. Large-area chips, however, are difficult to die-attach using gold eutectic die-attach methods, so we would use a tri-metallic alloy especially suited for soft-solder die attach to the header. This would *not* be recommended for high-frequency devices.

As we followed these several greatly simplified steps, illustrated in Fig. 3–2, we also identified in our example the doping concentrations for each step. Such doping concentrations are frequently plotted as shown in Fig. 3–3, which is called a *doping profile*.

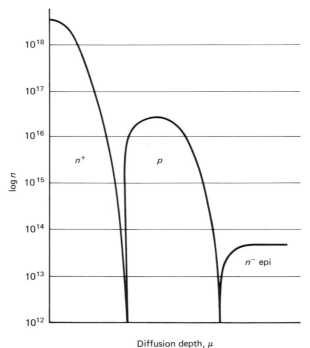

Figure 3.3 Typical doping profile, showing the reduction of doping concentration as we penetrate farther into the wafer. Between each *n* and *p* diffusion the profile must show *intrinsic*, that is, cross ϕ atom/cm^3 .

These fabrication steps that we have just concluded best describe the construction of a low-frequency power SIT which would be suit-able for either general industrial or audio applications.

We earlier identified the SIT as finding increasing popularity in high-frequency applications. If we are to succeed in the design and construction of a high-frequency power transistor we must be especially sensitive to the importance of its figure of merit. Remembering our discussion in Chap. 2, we must design the power FET to optimize both the forward transconductance and the input capacitance. The buried gate shown in Fig. 3–2 contributes to a high input capacitance that inhibits performance much above 5 MHz.

A novel design that shows great promise for very high frequency operation is shown in Fig. 3–4. Here the gate structure appears on the surface, where we have immediate ohmic contacts to metalization, thus achieving a drastic reduction of series gate resistance.

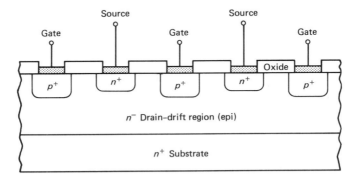

Figure 3.4   Cross-sectional view of a high-frequency static induction transistor.

We can, somewhat superficially, visualize the fabrication by following the steps in Fig. 3–5. The first two steps remain essentially unchanged from our previous example. The starting substrate material is again, n^+-doped silicon [Fig. 3–2(a)] with a doping level (for our example) of 5×10^{19} atoms/cm^3.

Since we are concerned with very high frequency performance the epi layer poses an interesting problem. If we wish to withstand a high breakdown voltage we need a lightly doped n^- layer of sufficient thickness to allow a fully depleted field. However, a new phenomenon to our study is an effect called *transit time*, which, like the triode vacuum tube, severely limits our upper frequency performance. As a consequence we find ourselves limited to a moderate drain–gate standoff voltage of, say, 100 V. Continuing our example, we might dope arsenic at approximately 1×10^{13} atoms/cm^3 into an epi layer 9×10^{-4} cm thick.

With the completion of our epi layer we come to a radical change from what we previously observed in the fabrication of the low-frequency industrial SIT. It is crucial now for us to monitor the gate area closely, for its size and proximity to the source will affect the input capacitance and the figure of merit of the transistor. Using suitable masks we now proceed to implant highly doped p^+ and n^+ regions as

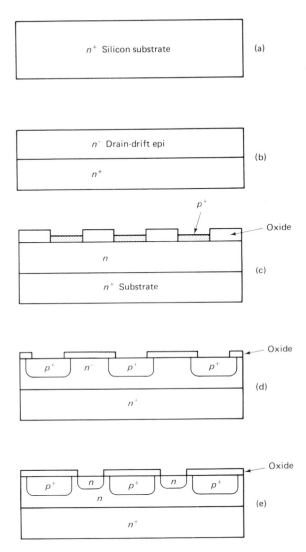

Figure 3.5 Basic construction steps involved in fabricating a high-frequency static induction transistor.

shown in Fig. 3-5(c). Continuing our example we could use boron for the p^+ with a concentration of typically 3×10^{19} atoms/cm^3 [Fig. 3-5(d)]. The n^+ would then be implanted after suitable oxidation and etching for the narrow window for arsenic deposition, which for our continuing example might have a concentration of 1×10^{18} atoms/cm^3 [Fig. 3-5(e)].

The remaining steps involve the laying of metal to pick up both the source (n^+) and gate (p^+) on the top surface and the drain from underneath. Since we are fabricating a high-frequency SIT, the metal for both the top and bottom must, of necessity, be a refactory metal involving gold. This is crucial for optimum high-power performance at high frequencies to alleviate what is commonly known as *metal migration*.

Nishizawa and others have shown a fundamental limitation in performance with this surface-gate structure. A further refinement has extended the performance of the SIT into the microwave region and has also improved the gain (although not necessarily simultaneously). A suggested cross-sectional view for a microwave SIT is shown in Fig. 3–6. Here the gate-to-source interelectrode parasitic capacitance is

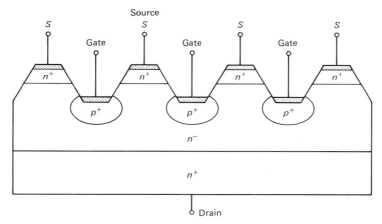

Figure 3.6 Cross-sectional view of a very high frequency static induction transistor where the gates are widely separated from the source metal to help reduce C_{in}.

further reduced by *channelizing* the gate structure. One precaution that we must not overlook is a slight increase in the source-to-gate channel resistance that could affect performance. A possible means to fabricate this channelized SIT would be very similar to the anistropically etched V-groove MOSFET, which we examine in detail later in this chapter. A scanning electron microscope (SEM) photograph of a channelized SIT is shown in Fig. 3–7.

3.3 THE VERTICAL DOUBLE-DIFFUSED MOSFET (VDMOS)

As we rethink what we read in Chap. 2 we remember that power FETs did not become popular until after the vertical short-channel MOSFET was introduced.

Figure 3.7 Scanning electron microscope view (SEM) of the very high frequency static induction transistor of Fig. 3.6, showing clearly the truncated gates. Each gate is 2 μm, wide and separation between gates is 5 μm.

We are able to appreciate the innovative skills of manufacturers as each tries to offer a device that outperforms its competitors. Generally, regardless of trade name, these power FETs all exhibit similar performance since their cell construction is basically the same. By cell we refer to a single channel extending from the source to the drain.

Among the plethora of short-channel power MOSFETs available, there are, however, two quite different structures. The first structure that we shall review is the vertical double-diffused MOSFET (VDMOS); the second is the V-groove MOSFET (VMOS).

Before we embark on our examination of the fabrication steps it is worthwhile that we compare some cross-sectional views of various DMOS structures. This we can do in Fig. 3–8, where a cross section of International Rectifier's HEXFET[1] and Siemens Aktiengesellshaft's SIPMOS[2] may be compared. Both devices are basically planar in that the n^+ implanted source, the overlaying oxide, and the silicon gate are all laterally disposed on the top surface. Aside from the obvious geometrical differences, the HEXFET being, as its name implies, a hexagonal structure; and the SIPMOS, rectangular, there is really little or no significant difference between them. They work the same way. As newer, catchier trade names proliferate within the family of MOSFETs, they will no doubt follow the fundamental short-channel concept. Thus, aside from some unforeseen and dramatic technological breakthrough of which we are presently ignorant, once we visualize the fabrication of a basic cell, we will understand them all.

We proceed as we did with the SIT by progressively following the fabrication steps of a basic cell, using Fig. 3–9 as our guide. We begin by choosing a silicon wafer, perhaps 76 to 100 mm in diameter and

[1] HEXFET is the trademark of the International Rectifier power MOSFET.
[2] SIPMOS is a trademark of the Siemens Corporation.

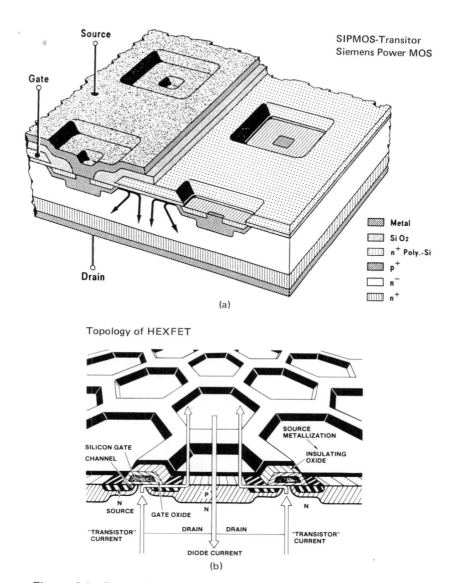

Source

Gate

SIPMOS-Transitor
Siemens Power MOS

Drain

Metal
Si O₂
n⁺ Poly.-Si
p⁺
n⁻
n⁺

(a)

Topology of HEXFET

SOURCE
METALLIZATION

SILICON GATE
CHANNEL

INSULATING
OXIDE

N
SOURCE

GATE OXIDE

P

N

"TRANSISTOR"
CURRENT

DRAIN

DRAIN

"TRANSISTOR"
CURRENT

DIODE CURRENT

(b)

Figure 3.8 Comparison of the Siemens SIPMOS and International Rectifier HEXFET. (SIPMOS is the trademark of Siemens Corp.; drawing courtesy of Siemens Corporation. HEXFET is the trademark of International Rectifier power MOSFET; drawing courtesy of International Rectifier, Inc.)

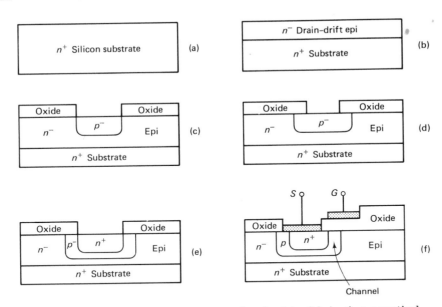

Figure 3.9 Basic construction steps involved in fabricating a vertical double-diffused MOSFET.

ranging from 0.4 to 0.9 mm thick, whose crystallographic plane is identified as <100>. A device physicist would recognize that although we would have the option of using any crystallographic orientation, <100> provides the optimum mobility, hence the optimum transconductance and lowest ON resistance. This substrate [Fig. 3-9(a)] is heavily doped n^+ with antimony. If we again use an example, we might dope this substrate to approximately 1×10^{18} atom/cm^3.

How we prepare and grow the epitaxy, [Fig. 3-9(b)] establishes both the breakdown voltage and the ON resistance. Both the doping concentration and thickness must be carefully controlled. Investigators have derived a relationship for the epi's contribution to the ON resistance,

$$R = \frac{\rho t}{A} f \qquad (3.1)$$

where ρ is a function of the doping concentration, t the epi thickness, and A the device's active area. The term f is based on the packing density of the complete active chip area. For our example we might dope with antimony to a concentration level of approximately 1×10^{15} atoms/cm^3.

Using suitable oxidizing, masking, and etching, we can prepare the epi surface to accept our first deep diffusion of boron (p) dopant [Fig. 3-9(c)]. Following our example we would start with a heavy *surface* concentration of, say, 1×10^{20} atoms/cm^3. This p-doped step is also a critical step in our processing, for it establishes the threshold voltage level of our MOSFET. The importance of a proper threshold will become more apparent as we continue in later chapters. We should be careful to note in Fig. 3-9(d) how, as we continue to diffuse this boron-doped stage, it *spreads out under the oxide*. On the heels of this boron implant we follow immediately *without modifying the oxide* with a very fast phosphorus, n^+ diffusion. By careful control we are able to arrive at a double-diffused section [Fig. 3-9(e)]. Thus we have what is commonly known as a double-diffused MOSFET.

Preparing the metal overlay is a bit more tedious than with the SIT. Again the manufacturer has the option of using an aluminum derivative or the more expensive, but more reliable, refractory metals. The oxide must be modified so that a proper ohmic contact will jump between the n^+ source diffusion and the p diffusion. The reason for this very important step will be fully examined in Chap. 4. So that we might invert the p *channel* we must have the gate metal (or the silicon-gate) overlay and extend beyond the p region [Fig. 3-9(f)]. Not only did we learn that the threshold voltage was controlled by the diffusion processing of the p-doped region, but we must also be aware that the gate design, including the oxide thickness, also controls the threshold voltage.

As with the SIT the back side of the chip is the drain contact and the manufacturer again has an option. For small-area devices a gold back is generally preferred, whereas for large-area and high-current FETs, a trimetallic suitable for soft soldering is best. In our fabrication of this VDMOS structure we doped both the substrate and the epi with antimony. This was obviously intentional. There are many high-temperature steps involved in the fabrication of a power FET, and antimony, being a highly stable dopant, is slow to change or migrate. That being so, we may rest assured that our initial processing steps will have remained as we had intended when the finished MOSFET appears. The source was doped with a highly volatile phosphorus because we needed a fast diffusion; but it was also our last high-temperature process.

A diffusion profile of our example is shown in Fig. 3-10. Note that the high surface concentration in the channel region has now diffused to a more reasonable level.

This example that we have just completed was, of course, an over-simplified design. Comparing Fig. 3-8 with Fig. 3-9, we should be quick to see that the former has a much different gate structure from our example. Our example, for simplicity, shows an all metal gate, whereas the two commercial VDMOSFETs both have what is called *silicon gates*. Silicon gates are made by a deposition of phosphorus-doped polysilicon embedded in silicon dioxide. This process is gener-

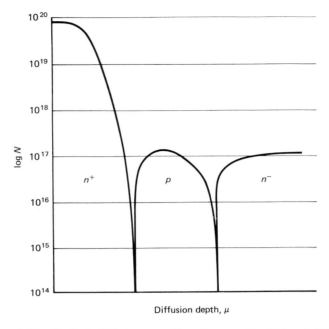

Figure 3.10 Typical diffusion profile of a double-diffused (DMOS) MOSFET.

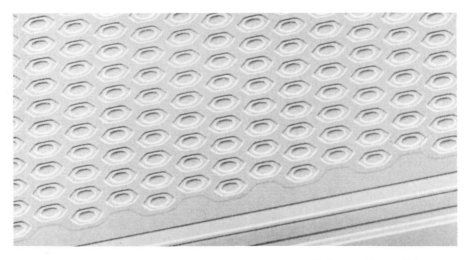

Figure 3.11 Scanning electron microscope view of the topology of the International Rectifier HEXFET. (Courtesy of International Rectifier, Inc.)

ally to be preferred to the metal gate process shown in our example. However, it does pose a problem, especially if we are interested in either high-frequency or high-speed switching applications. The problem is the silicon-gate's resistance: it is many thousands of times higher than a metal gate. A high series resistance gate plays havoc with our figure of merit [see Eqs. (2.2) through (2.5)].

The figure of merit can be greatly improved for silicon-gate structures by an overlay of metal integral with the silicon gate, but this still falls short of that which we can attain with the all-metal gate. We still find high-frequency, high-power VDMOSFETs with metal gates and probably always will.

A scanning electron microscope (SEM) photograph of a VDMOS-FET cell, shown in Fig. 3–11, closes our discussion on the fabrication of the vertical double-diffused MOSFET.

3.4 THE V-GROOVE MOSFET (VMOS)

We need not be particularly surprised to learn that the starting material is really no different than it was for both the SIT and the VDMOS. We will use Fig. 3–12 as we follow the fabrication steps of a basic VMOS structure. We begin, as before, with antimony-doped n^+ <100> crystalline orientation [3–12 (a)], upon which we grow an epi [3–12 (b)]. Both the substrate and epi are doped similarly to the substrate and epi we used for the VDMOS in the previous example. There are, however, some subtle differences that we should note as we compare Fig. 3–12 with Fig. 3–9. The inversion layer across the p channel for the DMOS

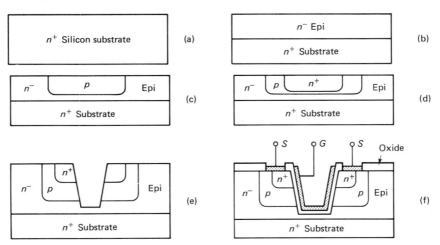

Figure 3.12 Basic construction steps involved in fabricating a V-groove VMOSFET.

structure is laterally placed near the uppermost surface of the FET, whereas for this VMOS structure the inversion layer (also across the *p* channel) penetrates angularly into the epi region, which effectively shortens the active epi thickness. Comparably built devices might show VDMOS with a higher breakdown voltage than with the VMOS, and VMOS with a lower ON resistance.

These early fabrication steps probably follow closer to that of a four-layer, high-beta bipolar transistor than do any of the power FET structures that we study in this book. Into the epi we first diffuse a deep boron-doped *p* well [Fig. 3-12(c)], followed by a phosphorus-doped n^+ well [Fig. 3-12(d)].

Upon the completion of this n^+ diffusion we have all but concluded the fabrication of a satisfactory bipolar transistor. The next few steps deviate quite drastically from the normal MOS construction which makes the VMOSFET unique.

After preparing the surface we immerse our wafer of "four-layer bipolar transistors" into an etching tank to produce the characteristic V-groove. This etching solution can be one of any number of chemicals, but we must be careful to choose a *nonisotropic* etchant. One suitable chemical for this task is potassium hydroxide (KOH). An anisotropic etchant will cease once the V-groove is complete, terminating in the $<111>$ crystalline plane. Early VMOS power FETs terminated the etch when the V reached a sharp point. However, because of the intense electric field, this sharp point developed within the n^- drift region, industry in general now terminates the etch so as to leave a truncated channel as shown in Fig. 3-12 (e).

Following this anisotropic etch we prepare our wafer by laying an oxide, being careful that no fine cracks appear about the edge of the channel. After masking and etching through the oxide as required, we evaporate metal—an aluminum alloy—for ohmic source contacts and the gate [Fig. 3-12(f)]. As we saw with the VDMOS structure, an alternative gate structure is frequently used called a polysilicon gate which would be embedded in silicon dioxide in the V-channel. As with the VMOS, the threshold voltage is affected by the oxide thickness over the *p*-doped channel. However, with VMOS the threshold voltage is also affected by the crystalline plane of the groove, the details of which will be reserved for collateral reading elsewhere. A scanning electron microscope (SEM) photograph in Fig. 3-13 clearly shows the unusual surface with the truncated V-grooves of a 400-V 8-A VMOSFET.

We are again reminded that this series of illustrations, depicting the fabrication of a VMOS power FET, are oversimplified. To ensure both high-speed switching and high-frequency performance care must be taken to reduce all possible parasitic capacitances, both within the structure itself as well as on the surface. Heavy metal overlapping over

Figure 3.13 Scanning electron microscope view of a truncated V-groove vertical VMOSFET. (Courtesy of Siliconix incorporated.)

areas where metal is not required for operation should be avoided. For example, in our fabrication illustration we would generally not wish for gate metal to lay over the source n^+ area, nor would we particularly care to have much exposure to the n^- drain–drift area, where high drain-to-gate feedback capacitance would result.

3.5 THE MESHED-GATE MOSFET

An unusual design is the meshed-gate vertical MOSFET. Unlike the previous power FETs that we have studied which were all n-channel, this FET is an enhancement-mode, p-channel MOSFET. That means, among other things, that we must impress a negative voltage on the gate to activate, or turn ON, the FET.

If, as we begin to study the fabrication of this MOSFET as outlined in Fig. 3-14, we give only a passing glance at its cross-sectional view, we may easily miss another unusual feature. What is most unusual with this MOSFET is that unlike those we have previously studied, this device achieves high current handling without relying on the short-channel design concept. In other words, it at first may *look like a DMOS structure*, but careful examination reveals that *it is not*.

As we discussed in Sec. 2.3, a p-channel device is the exact antithesis of the n-channel FET. Everything is opposite. Where for the n-channel we might have an n-doped region, for the p-channel we have a p-doped region. As we follow the illustrations in Fig. 3-14, we will develop the p-channel meshed-gate structure, but we could just as easily have reversed ourselves and develop an n-channel structure.

We begin fabricating this meshed-gate MOSFET by first growing an n epitaxy on a heavily doped p^+ substrate, [Fig. 3-14 (a)]. The inventors of this structure, Hitachi of Japan, suggest that we then dope this epi with approximately 4×10^{15} atoms/cm^3 of antimony [Fig. 3-14 (b)].

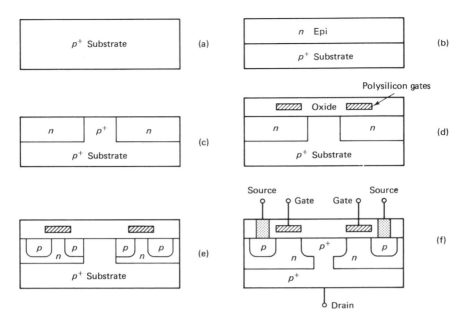

Figure 3.14 Basic construction steps involved in fabricating a meshed-gate MOSFET.

Our next step is a deep boron diffusion that, as we can see in Fig. 3-14 (c), extends a highly p^+-doped region through the epi into the p^+ substrate, forming both a surface drain and the interconnection to the substrate, which, for the final device, will be the drain contact.

Our next step in this fabrication cycle consists of growing an oxide overlay across the surface of the wafer and forming the polysilicon gate, as we have in Fig. 3-14(d).

To reduce both the gate-to-drain and gate-to-source capacitance, the next step is to first extend the drain right up to but not beyond the gate, and second to implant a source that also will not extend beyond the gate electrode. The best way to ensure this is to implant boron ions using *ion implantation*, a technique quite beyond the scope of this chapter. The result is a well-shaped p^+ region of the source and drain offering minimal parasitic capacitance between source to gate and gate to drain [Fig. 3-14 (e)].

Our final step is to metalize both the top surface to make ohmic contact with the source and the backside of the wafer for a drain contact [Fig. 3-14 (f)]. For the top surface we can use aluminum, but for the backside contact to the drain, we would generally use a gold eutectic for moderate-size geometries and a trimetallic backing suitable for soft-solder die attach for large geometries such as we might expect for this meshed-gate MOSFET.

Figure 3-15 shows a sketch of the meshed-gate MOSFET angle cut at 5°. Here we see that contact to the polysilicon gate can be at the perimeter of the chip. The deep p^+ wells appear as cylinders extending from the surface of the silicon (under the oxide) to the substrate. Surface metal contact to the source is through the oxide, as shown in Fig. 3-14(f).

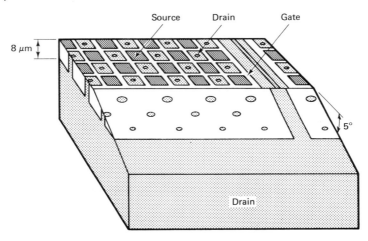

Figure 3.15 Sketch showing a cross-sectional view of a meshed-gate MOSFET. The cylindrical p^+ wells (looking much like the holes in Swiss cheese) connect the drain (substrate) to the surface drain contacts. [Courtesy of Central Research Laboratory, Hitachi, Ltd., Japan.]

Figure 3.16 Greatly enlarged photographic view of a single meshed-gate MOSFET chip. Here we see much of the surface devoted to source metal and a gate pad is situated at one corner. Having two large metalizations (the uppermost, the source, and the underneath the drain), this structure is ideal for handling heavy current. [Courtesy of Central Research Laboratory, Hitachi, Ltd., Japan.]

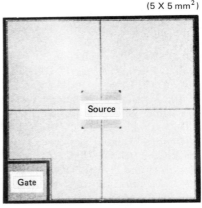

A scanning electron microscope (SEM) view of an actual chip that
has been angle-lapped resembling the sketch in Fig. 3-15 is shown in
Figs. 3-16 and 3-17.

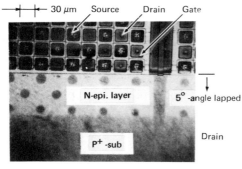

Angle-lapped

Figure 3.17 Scanning electron
microscope (SEM) view of the
meshed-gate MOSFET structure of
Figs. 3.15 and 3.16. [Courtesy
of Central Research Laboratory,
Hitachi Ltd., Japan.]

3.6 CONCLUSIONS

We have used this chapter to acquaint ourselves with various fabrication details of the more popular power FETs used today. We did
not try to elaborate on the details, and in particular we did not
develop the elaborate precautions required to improve the high-voltage
characteristics of the power FETs. By this we mean the field plates and
deep *p* wells generally used for controlling the depletion fields. We
leave these details for collateral study using the References that follow.

REFERENCES

AIGA, M., et al., "1 GHz 100 W Internally Matched Static Induction Transistor,"
   Conference Proceedings, 9th European Microwave Conference, Brighton,
   England, 1979, pp. 561-65.

BALIGA, G. JAYANT, "A High Gain Structure for Power Junction Gate Field Effect
   Transistors," *Technical Digest*, IEEE Electron Devices International Meeting,
   Washington, D.C., 1978, pp. 661-63.

COBBOLD, RICHARD S. C., *Theory and Application of Field Effect Transistors*.
   New York: John Wiley & Sons, Inc., 1970.

CSANKY, G., and J. BARTELT, "High Voltage S-Band Power Transistor," *First
   Interim Report*, Hughes Research Laboratory, Research and Development
   Technical Report DELET-79-0252-1, May 1980.

FUORS, DENNIS, and VERMA KRISHNA, "A Fully Implanted V-Groove Power
   MOSFET," *Technical Digest*, IEEE Electron Devices International Meeting,
   Washington, D.C., 1978, pp. 657-60.

GHANDI, S. K., *Semiconductor Power Devices*.   New York: Wiley-Interscience,
   1977.

HENG, T. M. S., et al., "Vertical Channel Metal-Oxide-Silicon Field Effect Transistor," *Final Report*, Westinghouse R&D Center, Pittsburg, Pa., Contract N00014-74-C-0012, Nov. 1, 1976.

HU, CHENMING, "A Parametric Study of Power MOSFETS," *Proceedings*, Power Electronics Specialists Conference (PESC), San Diego, Calif., June 1979.

JOHNSON, ROBERT J., and HELGE GRANBERG, "Design, Construction and Performance of High Power RF VMOS Devices," *Technical Digest*, IEEE Electron Devices International Meeting, Washington, D.C., 1979, pp. 93–96.

KAY, STEEVE, et al., "A New VMOS Power FET," *Technical Digest*, IEEE Electron Devices International Meeting, Washington, D.C., 1979, pp. 97–101.

LIDOW, A., et al., "Power Mosfet Technology," *Technical Digest*, IEEE Electron Devices International Meeting, Washington, D.C., 1979, pp. 79–83.

NAGATA, MINORU, "Power Handling Capability of MOSFET," *Proceedings*, 8th Conference (1976 International) on Solid State Devices, Tokyo, 1976.

NISHIZAWA, JUN-ICHI, "Field-Effect Transistor versus Analog Transistor (Static Induction Transistor)," *IEEE Trans. Electron Devices*, ED-22, No. 4 (Apr. 1975), 185–97.

NISHIZAWA, JUN-ICHI, "Recent Progress and Potential of S.I.T." *Proceedings*, 3rd International Conference on Solid State Devices, Tokyo, 1979.

NISHIZAWA, JUN-ICHI and KENJI YAMAMOTO, "High-Frequency High-Power Static Induction Transistor," *IEEE Trans. Electron Devices*, ED-25, No. 3 (Mar. 1978), 314–22.

OHMI, TADAHIRO, "Power Static Induction Transistor Technology," *Technical Digest*, IEEE Electron Devices International Meeting, Washington, D.C., 1979, pp. 84–87.

"Special Issue on Power MOS Devices," *IEEE Trans. Electron Devices*, ED-27, No.2 (Feb. 1980).

TIHANIGI, JENÖE, et al., MIS Field-Effect Transistor Having a Short Channel Length," U.S. Patent 4,190,850 (Feb. 26, 1980).

YUKIMOTO, YOSHINORI, et al., "1 GHz 20 W Static Induction Transistor," *Proceedings*, 9th International Conference on Solid State Devices, Tokyo, 1977.

four

Characterization and Modeling of Power FETs

4.1 INTRODUCTION

In this chapter our goal is to develop a more precise understanding of the performance of power FETs. We examine in some detail both their static and dynamic characteristics and attempt an analysis to determine how and why these characteristics occur. We also offer several perhaps too-simplistic models that will provide a clearer understanding of power FET performance.

Every data sheet we find is generally filled with data that are often either confusing or simply not sufficiently clear. All too often these data may seem irrelevant to the application for which we are seeking a solution by using a power FET. It is incumbent on the designer who plans to use a power FET first, to know how to select the right one by examining its characteristics, and second, to know how to use it properly.

There are two *basic* uses for power FETs. Either they are to be used as switches, where their sole responsibility is to turn power ON or OFF, or they are to be used as amplifiers.

As a switch we would expect to see very little power lost within the power FET itself other than what results from the intrinsic losses attributed to ON resistance. We would, of course, wish to keep these losses to a minimum.

As an amplifier we *expect* losses. These expected losses would appear as dissipative losses within the FET itself. Using a power FET as an active element in an amplifier, we would be especially conscious of the FET's *safe operating area* (SOA)—a term that we treat in detail later in this chapter.

There are yet other losses which, unless we were previously privy to, would go unrecognized, but not unnoticed. We shall identify and address these in this chapter.

No single model can define the power FET for all applications. If it were possible, it would be too complicated for us to use. To meet our needs in using power FETs either as switches or as amplifiers, we need several simple, albeit imperfect models. For example, if we are to build a switch we are especially interested in modeling the attenuation performance when the FET is OFF and its insertion loss when ON. Furthermore, we would want to understand the optimum gate drive mechanism. Such models would hardly suffice if we were to build a high-frequency amplifier. We would obviously need a different model.

Simple power MOSFET models such as shown in Fig. 4–1 poorly

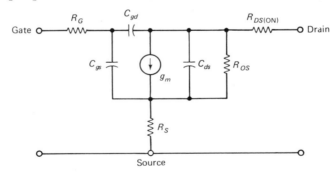

Figure 4.1 Simple electrical model of a MOSFET.

explain such phenomena as the characteristic nonlinearity in the sub-threshold region that we invariably witness when we view the transfer characteristics. Nonetheless, this simplistic model serves us well for simple two-port evaluations.

And so it goes—a reminder for us to view each model for the purpose for which it was developed. Regardless of the application of our power FET, we must know how to use it effectively and efficiently. Previous experience we have had with power bipolar transistors may have to be set aside. Power FETs operate on a different principle and we need to become familar with these principles before we can become proficient in using power FETs. We will have accomplished this when we conclude this chapter.

4.2 ELECTRICAL CHARACTERISTICS

Were we to examine any properly written power FET data sheet we would observe that its electrical characteristics are subdivided into two categories: *static* and *dynamic*. Static characteristics, as universally defined by the manufacturers, are the dc parameters such as breakdown voltage, gate leakage current, zero-bias drain current (I_{DSS}), and ON resistance. The dynamic characteristics are the small-signal ac parameters such as forward transconductance ($\Delta I_D / \Delta V_{GS}$), switching time ($t_{ON}$ and t_{OFF}), and capacitance.

Both static and dynamic characteristics play important roles depending upon the ultimate end use that we have chosen for the power FET. As a switch we would show particular attention to the static characteristics, in particular the OFF leakage currents and the ON resistance. If in our particular application we plan to hold off substantial voltages, then, of course, the breakdown voltage becomes an important parameter. If we were to use the power FET as an amplifier, the dynamic characteristics become particularly relevant. Transconductance and output conductance play important roles in establishing gain. Interelectrode capacitances which are voltage dependent become involved in the design of both high-speed switches and high-frequency amplifiers. Switching times are obviously critical for any high-speed application.

In the discussion to follow we concentrate our attention on several of the more important features, trying to reach a clearer understanding of the phenomenon attending each.

4.2.1 Breakdown Voltage

Although we are apt to believe that power FETs, as thermally degenerate devices, are generally quite forgiving when we perhaps momentarily stress them, breakdown may not necessarily be so. In this section we try to visualize the phenomemon that results in breakdown and the reasons why it may result in catastrophic destruction.

Our study, however, focuses on two quite different technologies. The J-FET—for which our attention has been upon the SIT—operates on quite different principles from those of the MOSFET. Yet the fundamentals of breakdown are similar. Regardless of our structure, the phenomenon we address first is commonly called *avalanche breakdown*.

To be sure we must be quick to point out that other forms of voltage breakdown exist. These forms of breakdown would be caused by the excessive expansion of the depletion fields surrounding the *p-n* barriers. Smart physicists, however, would configure the power FET design so that the first breakdown phenomenon is avalanche. There is a better chance for recovery if we approach and enter avalanche breakdown carefully.

Avalanche breakdown, as its name implies, is, indeed, an avalanche of carriers caused by the increasing electric field and increasing temperature. This increasing electric field, brought about by the increasing drain voltage, accelerates the majority (current) carriers. The pell-mell rush of carriers collide with electrons within the silicon crystalline lattice bumping them free to be, in turn, caught up by the accelerating electric field. So the process multiplies, or avalanches.

In Fig. 4-2 we see the effects of excessive drain voltage on a static induction transistor. The output characteristics show unsaturated drain

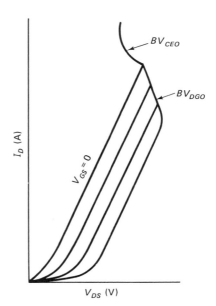

Figure 4.2 The breakdown phenomenon of the static induction transistor, where we see what appears as two independent effects. The first, as we would expect, BV_{DGO}, and then another quite unexpected breakdown that we might label BV_{CEO}.

current with increasing drain voltage until breakdown is reached. At that moment we witness what appears as drain–gate breakdown followed immediately by another perplexing phenomenon that has all the appearance of still another form of breakdown. In Fig. 2-18 we saw that the static induction transistor when positively biased assumed the performance of a *npn* bipolar transistor. Used in this mode the device was named the bipolar static induction transistor (BSIT). With this understanding of the SIT refreshed in our minds, let us return to that "other" form of breakdown shown in Fig. 4-2. What we may be witnessing is first the drain–gate breakdown followed by BV_{CEO}!

Avalanche breakdown is not the only breakdown mechanism. Other forms result from movement of the depletion fields within the transistor itself. These we examine below in some detail.

Two breakdown mechanisms resulting from expanding depletion fields are called *punch-through* breakdown and *reach-through* breakdown. If, in the MOSFET, we allow the drain voltage to increase sufficiently and the depletion field within the p-doped channel moves across to the n^+-doped source diffusion, current will flow and, in effect, bypass the regular channel. This phenomenon we call punch-through. It does no damage to the FET since it is merely bypassing the usually accepted current path.

Reach-through, on the other hand, occurs when the depletion field within the epitaxy "reaches through" to the substrate. Since it is this epitaxy that establishes our maximum operating voltage, any reach-through can be catastrophic. This type of breakdown would most certainly reflect on poor FET design.

Another very common breakdown mechanism whose cure has led to the successful design of many high-voltage transistors results from depletion-field *curvature*. We can easily reach a better understanding of this mechanism by using Fig. 4-3 to illustrate our discussion. Here

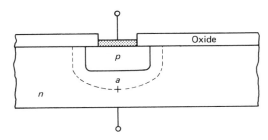

Figure 4.3 Depletion region about a simple p-n junction, showing potential breakdown at point a, where the depletion field makes its *closest approach*.

in this figure we can draw a depletion field about a simple p-n junction similar to those we saw in Chap. 2. We should note in this illustration that the depletion-field curvature is not equidistant from the junction but approaches closer to the barrier at point a. This should warn us that the maximum reverse breakdown voltage of this p-n junction is the breakdown voltage at a, the closest approach.

Another debilitating effect that grossly affects breakdown is the *crowding* of the depletion field, especially around corners, giving rise to highly concentrated electric fields that can result in premature breakdown.

From this we can draw some hasty but rational conclusions. It appears that if our breakdown is heavily dependent upon the curvature of the depletion fields, then either we exert a shaping influence on the field or we simply forget trying to produce high-voltage transistors!

For many years high-voltage transistor design resorted to a variety of depletion-field controls. These controls, formerly used to improve the breakdown characteristics of power bipolar transistors, are also useful in improving the breakdown performance of power FETs.

There are a variety of controls. Perhaps the best known is the use of *guard rings*. The two views in Fig. 4-4 graphically illustrate the marked improvement of adding guard rings to a high-voltage power MOSFET.

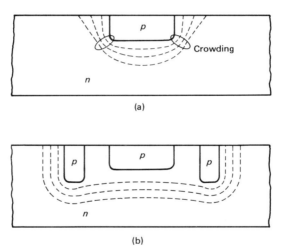

Figure 4.4   Exaggerated cross-sectional view of the simple *p-n* junction with and without guard rings. (a) Without guard rings crowding exists at corners, giving rise to potential breakdown. (b) With guard rings depletion fields are more evenly distributed with less opportunity for crowding.

Another control that has gained widespread attention is the *field plate termination*. Using a combination of guard rings and field plates, breakdown voltages exceeding 1000 V have resulted. In Fig. 4-5 we can compare the depletion field of a power FET made with and without a field plate termination.

Today all power FETs that reach beyond, say 250 V, use both guard rings and field plate terminations. By careful design we can achieve measured breakdown voltages caused by depletion-field move-

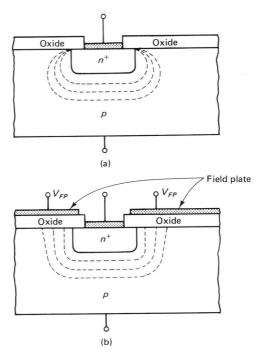

Figure 4.5 The effect of adding what is known as a field plate allows the fields to move more uniformly and lessens the possibility of a premature breakdown. (a) A *p-n* junction showing depletion-layer crowding when no field plate termination is used. (b) The effect of a field plate on the depletion layers.

ment better than 80% of the theoretical limit. Without these field-controlling measures, we would be fortunate to reach 50%.

4.2.2 Output Characteristics

Output characteristics provide important information regarding the relationship of drain current and drain voltage. For the family of power FETs that we are studying in this book we find two quite different relationships. For the static induction transistor the drain current is unsaturated when biased as a depletion-mode J-FET; for the vertical MOSFET we see, as we do for bipolar transistors, pronounced saturation. In Sec. 2.4.4 we explained Nishizawa's reasoning for this unsaturated drain current, so we need not repeat it here.

The power MOSFETs—VDMOS as well as VMOS—show current saturation with increasing drain voltage. Comparing Fig. 4–6(a) with 4–6(b) we would be hard-pressed to differentiate between them. Yet an observant reader will readily see that our power MOSFETs provide

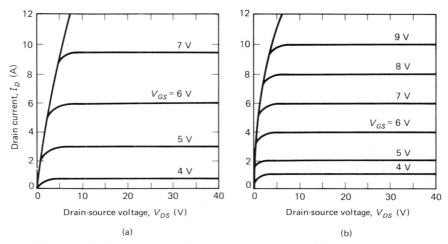

Figure 4.6 Comparison of the output characteristics of a typical DMOS (a) and a VMOS (b) power MOSFET of comparable ratings. These characteristics are seen to differ because of the differences in their transconductance [see Eq. (4-11)].

what amounts to limited *dual-mode* operation. At low drain voltages, as viewed in Fig. 4–6, our first mode is unsaturated drain current quickly rising to a sharp knee into the second mode, saturation. We can identify the first mode as a triodelike characteristic or a region of constant resistance; and the second mode as being a pentode. The mechanism that swings our MOSFET from the unsaturated region into the saturated region is caused by a phenomenon called *velocity saturation*.

Velocity saturation generally occurs in short-channel MOSFETs, where high electric fields predominate. Although the physics of velocity saturation is beyond the scope of this book, it is sufficient for us to acknowledge that both high forward transconductance and low channel ON resistance are directly affected. Furthermore, as we see in the following section, this mechanism gives rise to a linear transfer characteristic.

4.2.3 Transfer Characteristics

The transfer characteristics of a transistor provides us with an indication of how its output current follows the input drive. For a bipolar transistor the drive would be, in all likelihood, base current, I_B; for the vacuum tube, grid voltage, V_G; and for the FET, gate voltage, V_{GS}. The transfer functions of several devices may be mathematically expressed as follows.

Triode vacuum tube:

$$I_p = K_1\left(V_G + \frac{V_p}{\mu}\right)^{3/2} \tag{4.1}$$

Pentode vacuum tube:

$$I_p = K_2\left(V_G + \frac{V_{G2}}{\mu_{G2}}\right)^{3/2} \tag{4.2}$$

J-FET:

$$I_D = K_3\left(1 - \frac{V_{GS}}{V_p}\right)^2 \tag{4.3}$$

Bipolar transistor:

$$I_C = K_4\left(e^{\lambda\, V_{BE}} - 1\right) \tag{4.4}$$

Short-channel MOSFET:

$$I_D = K_5\left[V_{GS} - \left(V_{th} + \frac{V_{crit}}{2}\right)\right] \tag{4.5}$$

Static induction transistor:

$$I_D = \left(\frac{1}{K_6}\right)V_{DS} + \left(\frac{\mu}{K_6}\right)V_{GS} \tag{4.6}$$

Clarifying the terms found in these equations, first, we have a series of constants, K_1 through K_6, whose value is predicated on the physical design of the structure. For vacuum tubes, V_G represents the grid bias and together with the static induction transistor, which acts very much like a space-charge-limited triode, μ is the amplification factor. V_p is, of course, the plate voltage. For our short-channel MOS-FET we have V_{th}, the threshold voltage, which, as explained in Sec. 2.4.5.2, is the gate-to-source voltage at which point drain current commences to flow. The new term in Eq. (4.5) is V_{crit}, which is the drain voltage required to achieve velocity saturation. For short-channel MOS-FETs we can approximate this from a knowledge of channel length l in microns.

$$V_{crit} = lE_{crit} \tag{4.7}$$

Where E_{crit} has a nominal value of 2.5×10^4 V/cm.

We can construct a transfer characteristic chart from the output characteristics of an active device simply by plotting output current versus input voltage (or, for a bipolar, current). We can choose to plot either the static or dynamic characteristics. Earlier we understood that static implied dc, whereas dynamic meant ac. If we wish to construct the static transfer characteristics, we maintain a fixed V_{DS} and our graph will appear as shown in Fig. 4-7. If, on the other hand, we

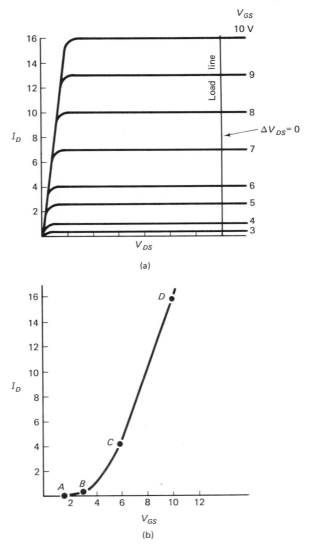

(a)

(b)

Figure 4.7 Derivation of *static* transfer characteristics from the power FET's output characteristics. A zero-resistance load line is equivalent to no change in V_{DS} (that is, V_{DS} = a fixed value). (a) Output characteristics with static load line superimposed. (b) Static transfer characteristics ($\Delta V_{DS} = 0$).

choose to plot the dynamic characteristics, we must first lay a *load line* over the output characteristics and from that we plot our graph. The latter would look like Fig. 4–8. It is important to understand that both

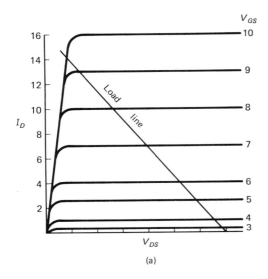

(a)

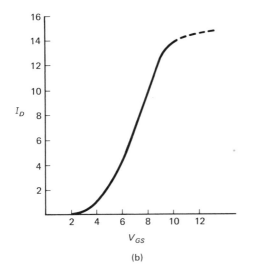

V_{GS}

(b)

Figure 4.8 Derivation of *dynamic* transfer characteristics from the power FET's output characteristics. Here using a finite load line we see what amounts to current crowding at the higher bias voltages, which causes the transfer characteristics to flatten at high bias levels. (a) Output characteristics with dynamic load line superimposed. (b) Dynamic transfer characteristics.

figures result from the same output characteristics. Our load line would simply represent the desired output resistance of our circuit.

If we examine the static characteristics, shown in Fig. 4–7, we see three distinct regions:

    A–B the subthreshold region

    B–C the square-law region

    C–D the constant g_m region where carrier velocity is reached.

In Fig. 4–8 we have yet another region of output current saturation resulting from the load line extending into the triode region of the output characteristics. This appears to suggest that at high drain currents and at high gate voltages the transconductance g_m is likely to decrease. This uppermost region of current saturation is probably dependent, as we might expect intuitively, on the value of the load line. As we reduce the load resistance we achieve increasingly greater operating range within the linear region. Were we considering using this MOSFET in an amplifier, we would see a proportional improvement in dynamic range and, at the same time, a decrease in overall stage gain.

When we speak of a transistor as being either linear or square-law or what have you, it is always with reference to the transfer characteristics. What we are looking for is the ratio of incremental cause and effect. When this ratio is constant, the device is linear.

4.2.4 Leakage Currents

If leakage must exist, we would expect to find it at either the gate or the drain. Generally, the symbols used for leakage are I_{GSS} for reverse-biased gate leakage and $I_{D(\text{off})}$ for drain current leakage when the power FET is biased OFF.

For the static induction transistor, gate leakage is simply the reverse-bias leakage of a *p-n* junction but only when the SIT is being used as a J-FET and *not as a BSIT*. This gate leakage may then be defined mathematically as

$$I = I_0 \left[\exp\left(\frac{qV}{kT}\right) - 1 \right] \qquad (4.8)$$

where I_0 is the reverse saturation current; q the electron charge (1.6×10^{-19} C); V the applied voltage, which in this case would be the voltage across the *p-n* junction; and the room-temperature value of kT is 0.0259 eV.

As the temperature increases we would expect to witness a proportional increase in gate leakage current according to Eq. (4.8). If we assume that the leakage phenomenon of the SIT behaves like its close family relation the planar J-FET, we would further expect to see the leakage current nearly double for every 11° increase in temperature.

OFF leakage current $I_{D(off)}$ is more difficult to define qualitatively for the SIT. With a high negative (reverse) bias on the gate, the depletion field has thoroughly choked off the drain current. In the equivalent electrical circuit, shown in Fig. 4-9, $R_{DS(off)}$ would be very high and $I_{D(off)}$ would, more than likely, consist mainly of drain–gate leakage current, I_{DG}.

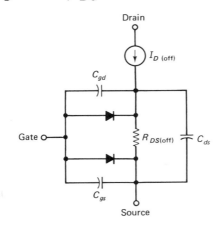

Figure 4.9 Equivalent OFF circuit for the static induction transistor.

Leakage current in our enhancement-mode power MOSFETs appears, qualitatively, to be limited to $I_{D(off)}$. This, of course, precludes that we have biased the MOSFET OFF. Universally, we see the MOSFET model showing the gate electrode as being perfectly isolated and hence exhibiting no gate leakage current. We find little information in the expansive literature, so we must conclude that the major source of gate leakage current in MOSFETs lies in either defective oxides or as a result of unclean processing. We must, however, be prepared to acknowledge that for zener-protected MOSFET gates, leakage currents not only can and do exist but can be predicted quantitatively!

If we believe that the power MOSFET of interest has zener gate protection, we need to become aware of a fundamental processing difficulty in the implantation of a true zener diode monolithically. By examining Fig. 4–10 we are suddenly made aware of the near impossibility of building a zener diode without effecting severe parasitic problems between it and the power FET. What we at first thought was simply a zener diode now becomes an intricately involved *npn* bipolar transistor whose behavior depends to a great measure on our biasing of the gate and source of the power MOSFET. Where we might at first think that a reverse-biased MOSFET gate would offer little or no gate leakage current, we now discover, perhaps to our amazement, that if we are not careful, destructive gate current is entirely possible! We set this problem aside for the moment while we continue with the various elec-

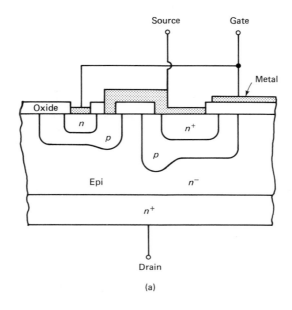

(a)

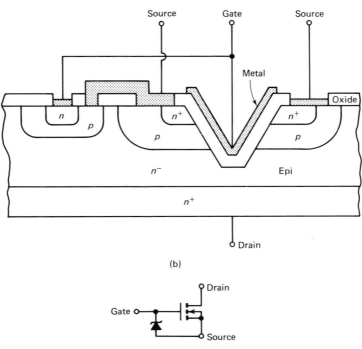

(b)

(c)

Figure 4.10 Cross-sectional view of both an *n*-channel DMOS (a) and a VMOS (b) power MOSFET showing the monolithic zener-protection diode.

trical characteristics of power FETs and return to the modeling of parasitic elements in Sec. 4.3.2.

Early in the development of power MOSFETs, OFF leakage currents were of a major concern. All too often, as a result of high temperatures, MOSFETs were found to possess certain instabilities that for the most part resulted in excessive drain leakage current. Where a newly fabricated power MOSFET might at first offer miniscule OFF drain current, during or immediately following a high temperature bake (often called "HTRB" or *high-temperature reverse-bias burn-in*), drain-to-source currents might be measured in the tens of milliamperes!

When we find ourselves faced with a high $I_{D(off)}$ leakage power MOSFET we can presume that our problem can be traced to one or more of three basic processing anomalies. We might have excessive ion drift in the gate oxide, possibly caused by some impurity such as sodium contamination; we may have trapped some ions on the silicon surface prior to growing the gate oxide. These free ions may, under high temperature, have worked their way into the silicon-doped diffusions; finally, insofar as this brief review will explore (and there may be others), we may be affected by a variety of what are called *surface-state* instabilities. More often than not, these various instabilities can be detected in the manufacture by what is known in the industry as a

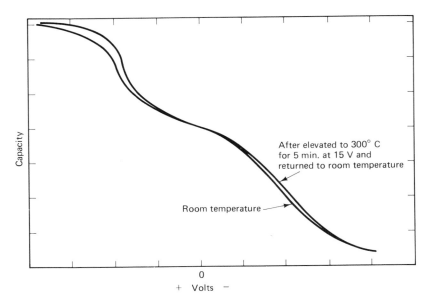

Figure 4.11 A typical *C-V* plot. One of the most useful measurements is the *C-V* plot. From this graph, dielectric constant, fixed surface states, and threshold voltages can be determined quantitatively if certain physical variables of the MOSFET are known.

C-V plot. A good C-V plot for a power MOSFET is shown in Fig. 4-11. The cure for excessive gate leakage and high $I_{D(\text{off})}$ currents is simply cleanliness in processing.

Before we close this section we should be aware of a problem that appears like $I_{D(\text{off})}$. This problem is particularly bothersome with power MOSFETs and pertains to the effects of increasing temperature on the threshold voltage. As the ambient temperature rises, the threshold voltage for all MOSFETs drops at approximately 5mV per $^\circ$C. Consequently, it is entirely possible for us to see an increase in drain current, I_D, that we might presume to be $I_{D(\text{off})}$. It all depends, of course, on how we biased the power MOSFET and how high we allow the temperature to rise. A high-threshold FET would show less sensitivity to temperature.

Summing up the problems of leakage, we would expect to see values of gate leakage for both the SIT and power MOSFETs well down in the picoampere range. The SIT would show the usual reverse diode current. If our power MOSFET has a zener-protected gate, our measured gate leakage current will depend on many factors. $I_{D(\text{off})}$ should also register very low values.

4.2.5 Interelectrode Parasitic Capacitances

Capacitance is a dynamic characteristic, as it is measured at a high frequency, generally 1 MHz. For any FET, Whether it is a SIT or a power MOSFET, we collect the various parasitic capacitances and identify them as

C_{iss}: input capacitance with drain and source shorted

C_{oss}: common-source output capacitance

C_{rss}: reverse transfer capacitance or gate-to-drain capacitance

Each of these capacitances are complex, consisting of numerous parasitic elements. Furthermore, these capacitances are voltage-dependent, rising to high values at low voltages and frequently decaying to quite small values at high voltages. In Chap. 2 we were acutely aware of the importance of low capacitance to achieve high figures of merit. If capacitance is inversely proportional to voltage, it seems reasonable that a high-voltage power FET would make a better high-speed switch or high-frequency amplifier than its physically identical but lower-voltage counterpart.

We find this voltage sensitivity to be due to the depletion fields that form the "plates" of our p-n junction capacitors. In Fig. 4-12 we can identify the various capacitances that make up C_{iss}, C_{oss}, and C_{rss}.

We can gain a rather sobering appreciation of the debilitating

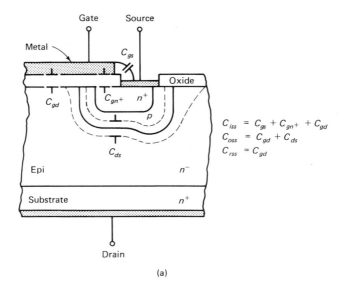

$$C_{iss} = C_{gs} + C_{gn+} + C_{gd}$$
$$C_{oss} = C_{gd} + C_{ds}$$
$$C_{rss} = C_{gd}$$

(a)

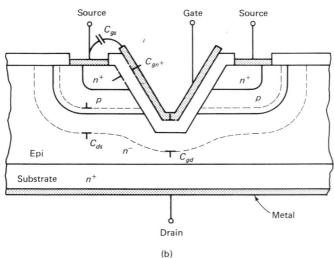

(b)

Figure 4.12 How C_{iss}, C_{oss}, and C_{rss} are formed in DMOS (a) and VMOS (b).

effects that parasitic capacitances play on power FETs if we simply reflect on one characteristic of the FET that we studied in Chap. 1 during our comparison of power FETs with other types of semiconductors. That characteristic was that the power FET exhibits no minority-carrier storage time. Without storage time we should have a very fast switch; in fact, our switch should be instantaneous, but it is not. Capacitance is the fundamental problem associated with performance limitation for FETs, whether they be power FETs or what have you. Many of the unusual designs that keep cropping up are simply novel ways to reduce the stray or parasitic capacitances and increase the figure of merit. There has been some excitement in fabricating power FETs on such nonconductive substrates as sapphire, simply to reduce these performance-limiting parasitic capacitances. The SIT that we saw in Fig. 3-7 offers a unique scheme for reducing C_{iss}. Double-diffused power MOSFETs reduce capacitance by restricting any metal overlays, in particular the gate metal over the source diffusion. V-groove power MOSFETs have resorted to an interesting fabrication process called shadow masking to reduce the input capacitance materially.

Although data sheets generally provide us with C_{iss}, C_{oss}, and C_{gss}, we must be aware that these collectively cover many parasitic capacitances. Were we to model a power FET, it would be somewhat awkward if we tried only to identify these three.

In Chap. 3 we mentioned that often power FETs can be constructed either for low-frequency industrial applications or for high-frequency applications. At that time we gave only passing mention of the problem of input capacity. Now let us turn our attention to the problem and potential cure of excessive input capacity.

Input capacity, generally identified as C_{iss} on most data sheets, includes, first, the gate-to-source capacity C_{gs}. For every power FET that we have considered, C_{gs} is more than just the capacity between the diffused junctions internal to the FET structure. We must also include what is called *field capacity*, the capacity between the surface metalization runs that link sources and gates. An additional source of input capacity depends upon the placement of these metalization runs as well as the placement of the bonding pads. If, for example, on one of our power MOSFET structures the gate metal overlays a source diffusion, we could expect an increase in input capacity. The reduction of field capacity is a major design effort in the original concept of the power FET.

The steps we take to lower the input capacity are the same steps to a successful design. To achieve this low input capacity we must focus our attention on the reduction of an insidious phenomenon called the *Miller effect*. Our success in reducing this effect has a greater impact on raising the figure of merit than any other factor in the design of a power FET, with the possible exception of transconductance.

The Miller effect is inherent in any voltage gain stage with a resistive load impedance where the active device—a transistor or a vacuum tube—exhibits a feedback capacitance between its input and output. We can identify a gain stage to be where the output signal appears across the drain resistor of the FET (or the collector of a bipolar transistor or the plate of a triode vacuum tube). A power FET operating common-source, that is, having the source common to both the input and the output, is prone to this debilitating Miller effect. What happens can be readily understood if we remember that the output signal of such a gain stage is 180° out of phase with the input. Feedback resulting from the gate-to-drain capacity C_{gd} effectively reduces the amplitude of the incoming signal so that this input essentially "sees" an effective capacity larger than what is actually present. We can resolve the Miller effect by the following mathematical expression:

$$C_{in} = C_{gs} + (1 - A_V)C_{gd} \qquad (4.9)$$

For us to gain an appreciation of the effect we need only follow a simple illustration. Let us consider a single gain stage using a power FET with the following characteristics:

$$\begin{array}{ll} C_{gs} & 35 \text{ pF} \\ C_{gd} & 6 \text{ pF} \\ g_m & 250 \text{ mmhos} \\ R_L & 200 \ \Omega \end{array}$$

The voltage gain can then be calculated:

$$A_V = 0.250 \times 200 = -50$$

(the negative sign representing 180° phase reversal between input and output). The Miller effect's contribution to the 35 pF of C_{gs} brings the total effective input capacity of the power FET to

$$C_{in} = 35 + [1 - (-50)] \, 6 = 341 \text{ pF}$$

It is pretty obvious that Miller effect can literally ruin the figure of merit of our gain stage and, furthermore, can effectively wipe out any high-speed switching if we do not take strong measures to overcome its effects. We discuss these measures when we get to Sec. 4.4. The "cure" for lowering the input capacity is to lower the feedback capacitance, C_{gd}.

Lowering the feedback capacity, C_{gd}, is not as easy as it might at first appear. For both the SIT and power MOSFETs, the first step would be to manipulate the n^- epitaxy both in doping concentration and thickness. Capacity is always a function of area, dielectric, and spacing; consequently, to reduce C_{gd} our task would be to manipulate all of these without compromising other important characteristics. A lighter-doped epitaxy would lower the capacity as would thickening the epitaxy. If this is feasible, then why isn't it done? The problem lies in yet another characteristic, which we describe later, called ON resistance. It would rise!

If we were to examine the capacity versus drain voltage characteristics for both the SIT and our power MOSFETs, we would see in all probability that as the drain voltage rose, the capacity would drop. The most pronounced capacity change is the feedback capacity C_{gd}. The change can be quite pronounced with some FETs offering as much as a 10 to 20 times reduction between 0 and 25 to 30 V. The change is especially noticeable with the SIT when pinch-off is passed, as shown in Fig. 4-13. This dramatic change results from the depletion fields about

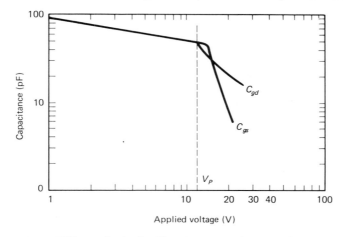

Figure 4.13   Effect of pinch-off voltage on C_{gs} and C_{dg} in a static induction transistor.

the gate structure literally "blooming" once pinch-off is reached. The effect on C_{gs} can also be quite pronounced. For the power MOSFET the effect is more subtle; however, for some power MOSFETs specifically designed for high-frequency operation as well as some small-signal MOSFETs, we see a similar characteristic once threshold is passed.

The feedback capacity of our double-diffused power MOSFETs (DMOS) can be greatly reduced by resorting to a self-aligning fabrication technique in the design. In the V-groove power MOSFET (VMOS) there are, perhaps, more options for us to reduce C_{gd}. Since the gate need only overlay the channel, we can truncate the V-groove, which we would probably do anyway to improve the breakdown, or we can increase the oxide thickness across the bottom of the groove, or we can increase the epitaxy doping and thickness with an attendant increase in ON resistance. We can call C_{gd} a depletion-dependent capacity, which makes the capacity dependent upon drain–source voltage. Increasing the voltage decreases the capacity. This capacity is made up of both the capacity between gate and substrate and the depletion region within the substrate in *series*. We could run a simple experiment to determine how much of this feedback capacity comes from the gate and substrate by reverse biasing the drain to source. Doing this we would completely collapse the depletion field, leaving only the gate-to-silicon substrate capacity. If we tried this we would have a far greater appreciation of the contribution the depletion field offers in reducing C_{gd} to manageable levels. We would also observe that the input capacity C_{gs} would not be as greatly affected.

We should be careful to understand that depletion-dependent capacity is critically affected by the movement of carriers. That is really what we mean when we say depletion-dependent. When either the DMOS or VMOS power FETs are operating so that their drain-to-source current flow is heavy, the feedback capacity is generally high. When we impress a high drain-to-source voltage on the drain and bias the MOSFET OFF, the feedback capacity is low.

Output capacity C_{oss} is also a depletion-dependent capacity whose value depends upon the impressed drain-to-source voltage. For the SIT this capacity is closely related to C_{gd}, and what we have discussed for C_{gd} holds reasonable well for C_{oss}. However, for the power MOSFETs, C_{oss} is the combined capacity of the drain-to-channel capacity, C_{ds}, and C_{gd}. Further into this chapter we will discover that the channel and source are electrically tied together as one element. If we were to plot the variation of C_{gd} and C_{oss} with drain-to-source voltage, we would see a relationship approximating (see Fig. 4-14)

$$C \approx \frac{1}{\sqrt{V_{DS}}} \qquad (4.10)$$

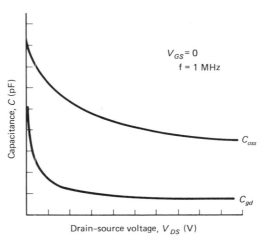

Figure 4.14   Effect of the Drain-source voltage V_{DS} on C_{gd} and C_{oss}.

When we developed the history of the static induction transistor in Chap. 2, we focused our attention on the development of high-transconductance FETs as we explored first the gridistor and then Zuleeg's MUCH-FET. In Fig. 2-6 we had conclusive proof of the superiority of the cylindrical J-FET over the conventional planar J-FET. Later we saw the cylindrical J-FET evolve into what we now identify as the static induction transistor.

With our power MOSFETs we have short-channel lengths that allow velocity saturation of the carriers to occur. This saturation is brought about by the high electric field extending from the drain potential. If the channel is short we also find current saturation as the drain voltage is raised. We will simply bypass the rigorous mathematical derivations for transconductance, which can be reviewed in the References concluding this chapter, and provide the standard formula as

$$g_m = \frac{\mu \epsilon_{ox} W (V_G - V_{TH})}{L T_{ox}} \tag{4.11}$$

where μ = effective mobility of electrons in the channel

ϵ_{ox} = oxide permittivity

T_{ox} = thickness of this oxide

W = width of the channel

L = length of the channel

V_{TH} = threshold voltage

V_G = applied gate voltage

Before we reach the erroneous conclusion that by forever shortening the channel, L will provide increasingly higher transconductance, we should be made aware that Eq. (4.11) is based on certain presuppositions as to the current saturation phenomenon. If we had chosen to undergo a rigorous derivation, we would have seen two limitations. One would be the eventual saturation of transconductance as the channel is shortened; the other would be an approaching independence of transconductance with V_G. Simply stated, we find that once we have applied a maximum gate voltage, it is quite useless to try to squeeze more performance with higher gate voltages.

It should be of interest to all of us to remember an additional aspect relative to the figure of merit that involves the channel length L. We might call this the maximum frequency, where our short-channel

MOSFET offers unity gain. This limit, which we would more properly label f_t, was given in Eq. (2.7).

If we return to Eq. (4.11), we can identify certain physical aspects that control the forward transconductance. The most obvious which appears to be a controllable variable is the ratio W/L as well as the oxide thickness T_{ox}. We have already concluded that a more rigorous examination of the derivations leading to this equation would preclude further adjustment of the gate voltage V_G. Intuitively, we should recognize that we do not wish to reduce the thickness of our oxide. If we did we would suffer a drastic lowering of threshold voltage V_{TH}, as well as possible premature drain–gate breakdown. So we are left with either further shortening of the channel, L, or increasing the channel width W. If we try further shortening of the channel we quickly approach the limits of our technology; if we widen the channel instead we simply increase the parasitic capacitance C_{gs}. Again, we are faced with a compromise. Yet what would happen if we opted to drop V_{TH}? Later we will discover that a low threshold voltage is not desirable, and we will wait until then to discuss why.

There is one aspect of transconductance that we did not mention: the effect of parasitic *resistance* on transconductance. Equation (4.11) assumes that no parasitic resistance exists, and for the ideal short-channel MOSFET we can anticipate that measured transconductance would closely follow Eq. (4.11). But if it does not, can we resolve the difference? Nishizawa recognized the problem, which led to his static induction transistor. Figure 2–17 identifies this parasitic resistance as providing negative feedback. We can express the effects of this parasitic resistor mathematically:

$$g_m' = \frac{g_m}{1 + g_m R_s} \qquad (4.12)$$

The value of R_s is not ON resistance per se, but only a very small portion of the total ON resistance of the MOSFET. Later in this chapter when we examine the physical model we discuss R_s somewhat more fully.

4.2.7 ON Resistance

When power MOSFETs were introduced they met with considerable resistance in the marketplace simply because when compared to the bipolar transistor that they were *supposed* to replace, they had unacceptably high V_{SAT}, or voltage drop under full conduction. Our problem was simple to recognize but not so simple to resolve. Nishizawa's static induction transistor did not have the same problem.

Right from the start, having recognized the fundamental limitations of the planar FET, the SIT was designed for minimum ON resistance. Time will tell which technology will provide the lowest value, the SIT or the short-channel power MOSFET.

When the short-channel power MOSFET was introduced in the mid-1970s we had no choice but the V-groove MOSFET (VMOS). Production quantities of the double-diffused MOSFET were unavailable until early in 1980. For the most part the problems of ON resistance are much the same for either the VMOS or the DMOSFET. There are subtlties and we address them in this discussion.

To begin we must gain an appreciation of the basic problem. Earlier we reviewed the problems of breakdown voltage and concluded that we could increase breakdown voltage at the expense of increasing the resistance within the n^- drain–drift region. We can approximate the relationship with the following equation:

$$R_d = kBV^{2.2 \text{ to } 2.7} \tag{4.13}$$

ON resistance is the sum of a rather lengthy series of parasitic resistances, including R_d, the drain–drift epitaxy resistance as given above. We could begin with the metalization runs of source metal, then the n^+ source diffusion, the channel itself when it is inverted due to a positive gate potential (for an n-channel MOSFET), plus the important drain-drift region and the substrate. Although with our vertical power MOSFETs the backside of the chip is the drain, we must still reckon with the problems of die attach as adding some resistance. Finally, we have the package itself. Even the pins of the power package, TO-3, provide measurable resistance! We can string these discrete resistive elements into an equation that would look something like this:

$$R_{ON} = R_{run} + R_{contact} + R_{n^+} + R_{channel} + R_{drift} + R_{sub} + R_{attach} \tag{4.14}$$

Of course for our discussion we can forget some of the lesser parasitic elements. The important elements for us to consider are the channel resistance $R_{channel}$ and the drain-drift resistance R_{drift}. Channel resistance is, as we might expect, dependent on gate voltage. It is easy for us to relate to this, as we generally think of FETs as being *voltage-variable resistors* (VCRs) whose channel resistance is a function of the gate-to-source bias. A power MOSFET is not much different. In fact, as we can identify the ON resistance of a J-FET by observing its output characteristics, so we can measure the ON resistance of a power MOSFET by its "triode" region within the output characteristics as shown in Fig. 4–6. While we are in this triode region, Ohm's law prevails and it becomes an easy matter to calculate the static ON resistance.

Of all the elements we see in Eq. (4.14), R_{drift} is by far the singularly most important. We have already witnessed its relationship to the breakdown voltage [Eq. (4.13)], but we should recognize that it is also dependent upon many other factors other than epi resistivity.

During our discussion of the basics of fabrication in Chap. 3, we barely grazed the problem of the epitaxy's contribution to ON resistance. In fact, we established a simple equation in Eq. (3.1) but left the reader with what we called a *packing density* term. Now we turn our attention to the effects of packing density and its contribution to that rather simplistic equation. Perhaps by now the reader realizes that to achieve both high current and low ON resistance we have to parallel many active cells, each cell consisting of at least one source, channel, and drain region.

In Chap. 3 as we followed the fabrication details of the vertical V-groove power MOSFET (VMOS), we somewhat offhandedly mentioned that industry no longer etches the groove to a sharp point but truncates the groove "so as to reduce the intense electric fields." This is, of course, true; but if we turn our attention to Fig. 4-15 we can see how currents spread out in the epitaxy region. Using the most general terms we can assume that current spreading will run 45° symmetrically about the point of entry of current into the epitaxy. Current spreading is further complicated by the close proximity of the *p-n* junction which comprises the channel–epitaxy region. Many investigators begin with several assumptions and proceed from there. Since we are not committed to the study of semiconductor physics we will approach this problem phenomenologically; that is, how does it affect what we want to do?

The effect becomes obvious if we draw several "cells" with a high packing density so as to "improve" our total current-handling capability, or so we might hope. Figure 4-16 illustrates our exaggerated densely packaged VMOSFET. The problem should be obvious. If it is not at first, we must be conscious that a semiconductor such as silicon has a maximum current density for optimum performance. What we have shown in Fig. 4-16 is a power VMOSFET that would not offer the user optimum performance.

The same problem that we have illustrated for the VMOS also holds true for the VDMOS. Packing density must be carefully considered by the manufacturer of power MOSFETs. There are limits in design, and as a consequence higher and higher drain currents must use larger geometries.

According to some investigators our problem may be compounded beyond what Figs. 4-15 and 4-16 show. Some believe that there is a far more debilitating effect that may result if we try to increase the current density of our MOSFET chip through the simple expedient of

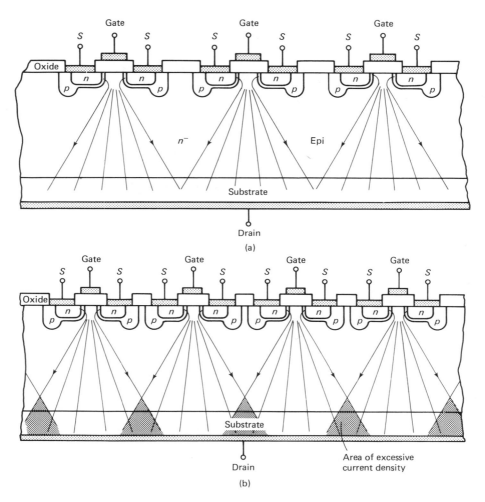

Figure 4.15 Effect of packing density on current flow in a vertical DMOS power MOSFET.

tightening the packing density. As we studied in Chap. 2, we are aware that a *depletion region* extends beyond the p-doped well in an n-channel FET (and beyond the n-doped well in a p-channel FET) regardless of whether we are considering a J-FET or a MOSFET. In Figs. 2–2 and 2–7 and the accompanying discussion, we saw that the drain–source voltage V_{DS} directly affects the pinch-off voltage V_p. If we now reconsider our cross-sectional view shown in Fig. 4–15 redrawn as we have done in Fig. 4–17, we see a depletion region extending out from the p wells (for an n-channel MOSFET). If we are to follow the reasoning of these investigators, we need but little imagination to discern a potential

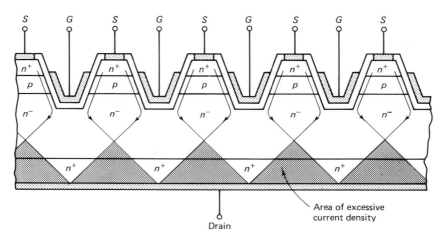

Figure 4.16 Effect of excessive packing density on current flow in a V-groove VMOS power MOSFET.

problem that can seriously inhibit the performance of a tightly packed, high-current MOSFET.

By following the reasoning of these investigators, if we now examine Fig. 4-17, we see that as our drain-source/body voltage increases, this depletion region begins to balloon *directly into the path of our drain current!* An equivalent circuit showing this presumed undesirable effect would show a J-FET *in series* with the MOSFET drain, as we have shown in Fig. 4-18.

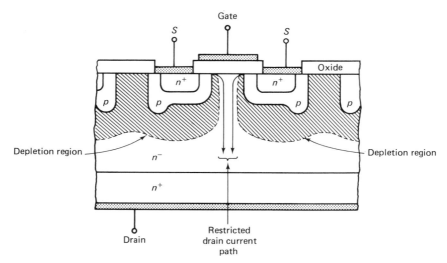

Figure 4.17 Effect of spreading depletion regions upon the drain current which results from too tight packing.

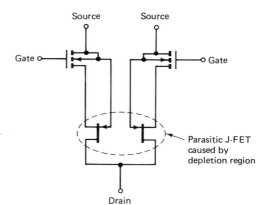

Figure 4.18 Simplified equivalent circuit of a too tightly packed power DMOSFET, showing the parasitic series-connected J-FET restricting drain current at high drain-source voltages.

However, we find other investigators who decry this hypothesis since, as they claim, a high drain–source voltage exists only at *low* drain currents, where the effect, if existing, would be muted. As our drain current increases, the drain–source voltage $V_{DS(\text{on})}$ decreases, since high currents force a proportionally higher voltage drop across our load rather than across the MOSFET. Consequently, these investigators claim that the depletion region *shrinks*.

We should perhaps be aware that it is entirely possible to have a power MOSFET stressed with high voltages simultaneously with high drain currents. We study specific situations later in this book. But for now we might consider a situation where we have a power factor that either leads or lags, causing, during switching, both a high voltage and a high current to exist across and through our FET simultaneously. Under such operating conditions we might, indeed, witness current limiting caused by the depletion crowding of the drain current through the drift region of our power MOSFET. This phenomenon may be difficult to detect, and we may have inadvertently covered the problem by providing on the data sheet a safe operating area, which we discuss later in the chapter.

In our discussion of ON resistance we concentrated rather heavily on the power MOSFETs and paid scant attention to the SIT. Unfortunately, the ON resistance of this power J-FET is rather poorly understood. Measurements show a dependence of $R_{DS(\text{on})}$ on both the gate-to-source voltage (as we would expect) *and the drain current*. As we increase the drain current, we see a lowering of ON resistance! This would suggest that if we were to use the SIT as a switching element, say, in an analog switch application, ON-resistance modulation could result. This would not be especially desirable. However, because of the unusual behavior of the SIT when we apply positive gate bias, this device might make a very interesting switch despite the potential of channel ($R_{DS(\text{on})}$) modulation.

The subject of ON resistance is endless. Before we close our discussion we should be made aware that as our $R_{DS(on)}$ changes with bias, we must watch that we do not exceed the dissipation of our device, remembering that watts equals I^2R. Finally, we must remember that for a unit area geometry, that is, for a power MOSFET of prescribed area, breakdown voltage ultimately defines the ON resistance of the device.

Our next topic in this section should be on the very important subject of safe operating area (SOA). No engineer versed in power transistors ever forgives a manufacturer who fails to clearly identify the safe operating area of their transistors. Yet because of the mechanics of SOA for the power FET, we first examine the models and then conclude with a discussion of SOA.

4.3 THE ELECTRICAL AND PARASITIC MODELS

4.3.1 The SIT

In earlier chapters we saw that the evolution of the static induction transistor (SIT) was propelled by a desire to solve the degenerative property that is common to all planar J-FETs. By reducing the intrinsic source resistance the feedback was sufficiently attenuated to prevent current saturation. As we read in Sec. 4.2.7, although the SIT closely resembles a J-FET, it has yet to be fully understood, especially with regard to the current-ON resistance phenomenon. A J-FET model that resembles the SIT is shown in Fig. 4–19. This electrical equivalent circuit typifies *any* J-FET and is not unique to the SIT.

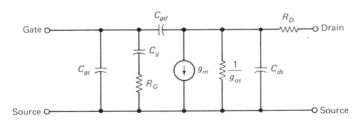

Figure 4.19   Electrical model of the static induction transistor.

4.3.2 DMOS and VMOS

We have chosen to examine these two power MOSFETs together because of their close relationship. Power MOSFETs have enjoyed a far more rigorous examination and several models exist that can aid us in understanding their performance. There are several parasitic elements that are common to both the DMOS and the VMOS power

FETs. These parasitic elements were discussed briefly in Chap. 2. All but one of these elements is intrinsic to the basic structure, and as we shall discover, our control over these elements plays an important role in establishing the usefulness of the power FET.

All too often we associate parasitic elements with something unwanted; perhaps that is why we frequently label such elements as *parasitics*. Yet there is one "parasitic" element common to both MOSFETs that is generally more beneficial than otherwise, and that is the *p-n* diode between the *p*-channel and the n^- epitaxy shown in Fig. 4–20 (and earlier in Fig. 2–26). Some have said, and we do not disagree, that this reverse *p-n* diode is a hidden bonus for the circuit designer. That statement may be a bit too strong, for we will see one disadvantage in Chap. 9.

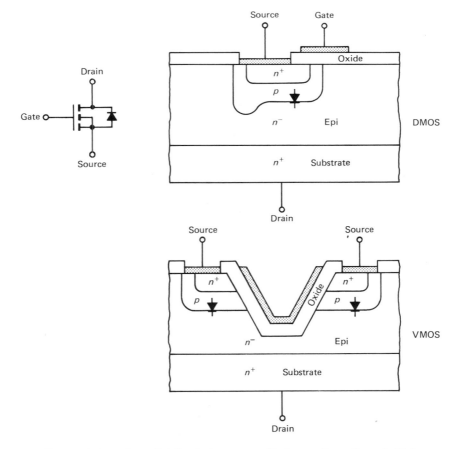

Figure 4.20 Beneficial" drain–source diode resulting from bridging the source to body of both the DMOS and VMOS structures.

There is nothing particularly unusual about this *p-n* diode as diodes go. Fortunately for the operation of the power MOSFET, it is *reverse*-biased and does contribute to the OFF leakage current between the drain and source as would any reverse-biased diode placed across a unipolar switch. Since we do have this diode in shunt with our power MOSFET, it does suggest certain questions that need answers. In this chapter we do not discuss how this diode finds useful applications; we reserve that for later chapters. Our questions concern its characteristics, such as current capacity in the forward-biased mode, how fast the diode recovers in a switch-mode configuration, and when conducting, what voltage drop we might expect across the diode.

To answer these questions we should look first at this parasitic diode alone, free of the encumbrances of the shunting MOSFET. This we can do by using Fig. 4-21 as our model. Diodes are often classified

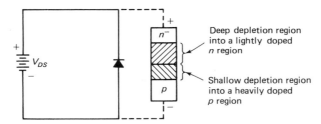

Figure 4.21 Model of the parasitic drain–body diode, showing that for an n^--p junction the depletion layer is greatest within the n^- region.

as being either abrupt or diffused junction devices. The diode we have in these power MOSFETs cannot be positively identified as either, but if we had to choose between them we would say that it is more closely related to the abrupt junction. An abrupt junction diode often has one side of the junction heavily doped in comparison to the other side, which makes a very asymmetrical structure. That pretty much identifies this diode. The *p*-doped channel is heavily doped, whereas the n^- drift region is rather lightly doped. This is somewhat contrary to the usual abrupt junction diode, which would have a heavily doped n region and a lighter doped *p* region. As we study the parasitic effects of this diode we must be sure to keep in mind that under normal operating conditions the diode "operates" under reverse-biased conditions. In certain applications, which we address later in the book, the diode plays an important role during its forward conduction mode.

Earlier in this chapter we studied various forms of breakdown phenomena, one of which was identified as reach-through breakdown. Since our diode relies on the same doping profile as the parent MOSFET, it is altogether reasonable to believe that the reverse breakdown potential for this diode equals that for the MOSFET; and so it does.

How much current this diode can support in the forward conduction mode is somewhat academic since it is formed in the parent MOSFET. When the MOSFET is biased full ON, its current capacity is determined by several factors, such as its thermal resistance (dissipation), the size of the bonding wires (how much current they can pass), the capacity of the metalization runs laid down on the chip surface, and, of course, the SOA of the FET itself. The same boundaries exist for the diode. Possibly the major limitation is the thermal resistance of the chip.

When we began this book we were soon led to believe that power FETs, like any FET, were totally devoid of minority carriers. That is perfectly true; all FETs are majority-carrier transistors and do not rely in any way upon minority-carrier injection. But now we are going to discuss the role of minority carriers in the performance of this parasitic diode. Our original statement is still true because the minority carriers are not a part of the FET proper but are associated with the parasitic element—the diode.

In the applications that we discuss in later chapters, where we find the usefulness of this reverse-biased *p-n* diode used to the fullest, we need to have some understanding regarding its *reverse recovery time*.

Reverse recovery is simple to grasp if we remember basic diode conduction theory. In the forward conduction mode with a positive voltage on the p and a negative potential on the n^- epitaxy, the depletion fields are totally collapsed and current is freely flowing across the junction. If we were abruptly to reverse the polarity across the diode, it takes a finite amount of time for the depletion fields to become reestablished. Carriers that were previously moving are now swept out of the diode as the depletion field builds. These "misplaced" carriers are now called minority carriers, and the time it takes for them to be removed is what is commonly called the reverse recovery time. What is of special importance to us is that until these minority carriers are removed, current continues to flow in the forward direction despite the fact that the diode is reverse-biased. The effect is easily visualized in Fig. 4-22. Fortunately, most all of the power MOSFETs on the market today have rather swift reverse recovery times, but for those applications where ultra-high-speed switching is important, it would be well to examine the particular MOSFET under consideration. The fundamental problem that we need to be aware of is that if we were to switch the power MOSFET across a dc line and the diode was not fully recovered, we would have what amounts to a short-circuit or at least a low-resistance path. The current would soar and if the transistor was not properly mounted to a good heat sink, it could conceivably suffer irrepairable damage.

The forward voltage drop across the parasitic diode is a function

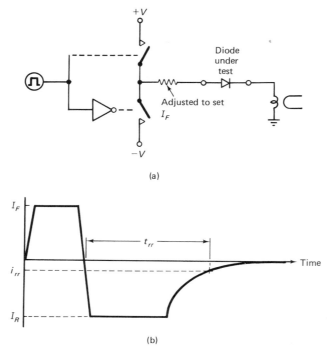

Figure 4.22   Reverse recovery-time test circuit (a) and resulting waveforms (b).

of Ohm's law and the gate bias on the MOSFET. In Fig. 4–20 we saw that the parasitic diode *shunts* the power MOSFET; consequently, any current flow through the channel as a result of FET action effectively reduces the voltage dropped across the *p-n* diode. If, however, we have no gate bias, we can expect the forward voltage to be similar to what we might expect from a silicon rectifier. The value generally falls somewhere from 0.6 V to less than 2 V for a high-voltage power FET. A well-prepared commercial data sheet should provide these characteristics.

The second parasitic element intrinsic to the power MOSFET is not beneficial under any circumstances, except possibly for the SIT. A closer look at Fig. 4–20 reveals that both the *n*-channel DMOS and VMOS have within their structure a parasitic *npn* bipolar transistor (a *p*-channel MOSFET would have a *pnp* bipolar transistor). Furthermore, we see that this parasitic bipolar transistor is in shunt with the MOSFET, which means that if, for any reason, we inadvertently activate this bipolar transistor, we will effectively cancel any beneficial features of the power FET! Our whole purpose in using power FETs was to rid ourselves of the deleterious bipolar transistor problem. We surely do not want them back in disguise as a FET.

Before we continue with the power MOSFETs, let us turn our attention for a moment to the SIT. We stated that the parasitic bipolar *might* be beneficial to the SIT. In our study of the SIT in Chap. 2 we discovered a most unusual feature, in that when a negative gate bias is applied, the device acts like a space-charge-limited triode, but when we place a positive bias on the gate, the SIT appears to revert into a bipolar transistor. Operation in this mode, as we saw in Chap. 2, identified the device as a BSIT—a *bipolar static induction transistor*.

For the power MOSFET we can begin to visualize the electrical equivalent circuit of both the parasitic *p-n* diode and the parasitic *npn* bipolar transistor. In Fig. 4–23 we see that the base of this parasitic

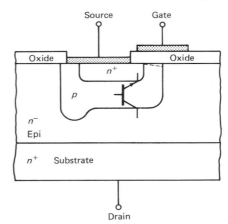

Figure 4.23 Double-diffused structure showing that the base of the *npn* parasitic bipolar transistor is common to the MOSFET channel.

bipolar transistor is common to the channel of the MOSFET, the emitter is common to the n^+ source, and the collector is common to the n^- drain–drift region. In Chap. 2 we discussed in some detail the mechanics of operation of this parasitic bipolar transistor and we learned why all power MOSFETs have their source and channel diffusions electrically bonded together. Yet as we also discovered in Chap. 2, bonding the source and channel did not guarantee relief.

The problem is not impossible to resolve, and indeed manufacturers have solved it. The parasitic bipolar transistor is not unique to power MOSFETs. For years CMOS circuitry had been plagued with a similar problem that at one time caused a phenomenon called *latchup*. The cure to rid ourselves of this parasitic bipolar transistor is to either render it inoperable or to so reduce its effectiveness that it no longer becomes menacing. As we saw in Chap. 2, manufacturers of power MOSFETs generally tie a bridge between the source and the *p*-doped channel using some sophisticated methods to ensure complete muting of the parasitic bipolar transistor.

Although we may have succeeded in muting the action of this parasitic *npn* bipolar transistor, we must not be lulled into thinking that the parasitic *elements* of this transistor do not affect performance. In the concluding section of this chapter we examine a model and review the consequences that we would suffer by not taking these elements into consideration.

Some manufacturers of moderate-power MOSFETs, say, with current ratings of 2 A or less, have supplied what appears to be zener-gate protection, ostensibly to protect the gate oxide from static discharges that might result from handling. This so-called zener can become our Achilles' heel if we are not aware of its parasitic nature. Although the electrical circuit of this zener is quite straightforward, as shown in Fig. 4-24, placing it on a monolithic vertical MOSFET is quite another problem, as we can see in Fig. 4-25. We can at once see that

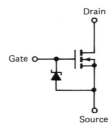

Figure 4.24 Electrical schematic of a zener-protected gate MOSFET.

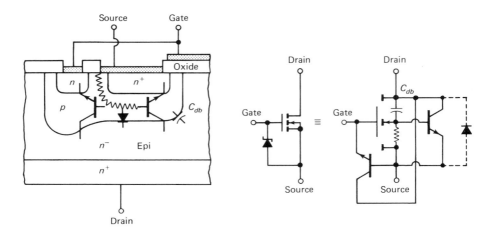

Figure 4.25 Total parasitic equivalent circuit of a zener-protected vertical power MOSFET, again showing (*for clarity sake only*) the body–collector diode separately from the parasitic *npn* bipolar transistor.

we have another parasitic *npn* bipolar transistor. In this case the parasitic bipolar transistor's emitter is tied to the gate of the MOSFET, its base is intrinsic to the channel, and its collector is formed by the n^- drain–drift epitaxy.

It is easy to see how this zener can be a subtle problem in some applications. To turn the *n*-channel enhancement-mode power MOS-FET ON, we positively bias the gate. With the zener back-biased we have no problem, nor do we have a problem with the parasitic *npn* bipolar transistor formed by this monolithic zener. But what happens if we swing the gate potential negative? We have, in effect, raised the base voltage of the parasitic bipolar transistor and thus activated bipolar action. Now with this parasitic bipolar transistor activated, we have effectively shorted the MOSFET's drain to the gate. If, for example, we were using a moderately high voltage zener-protected MOSFET, we might accidentally have punctured the gate oxide. Were this to occur we could say that we have destroyed the MOSFET.

4.4 MODELING FOR SWITCHING

Earlier in this chapter we discussed ON resistance in some detail. Here we shall examine the mechanics of turning the power FET either ON or OFF. Turning the power FET either ON or OFF requires us to examine its dynamic input characteristics. The problem we face in switching is one involving the parasitic capacitances. They must be considered if we intend to benefit from the FET's inherently fast switching speed. Our focus will be principally on two capacitances: the input capacity C_{gs} and the drain–gate feedback capacity C_{gd}.

Earlier in this chapter we examined the various parasitic capacitances and concluded that any switching speed less than instantaneous was simply due to these capacitances. We now try to better acquaint ourselves as to how these capacitances affect switching speed.

We know that the dc input resistance of either a J-FET or our power MOSFETs is very high. For the MOSFET the gate is entirely insulated with silicon dioxide and a resistance measurement might easily read in the billions of ohms. The dynamic input impedance is quite another story and depends directly upon capacity, the frequency, or its reciprocal, time.

We also understand that if we wish to build a charge across a capacitor, it will require a finite amount of energy. In other words, charging a capacitor takes current.

When we first turn ON a power FET we start charging the input capacitor. For the MOSFETs that have a finite threshold voltage, we can easily calculate the charging current using the simple equation

$$i = C \frac{\partial V}{\partial t} \tag{4.15}$$

In some respects we are fortunate that most of C_{gs} is either field capacity or capacity between the gate and the n^+ source. Capacity between the gate to channel and gate to epitaxy (which would be C_{gd}) is dependent upon the gate voltage as well as the drain, or channel current. If the preponderance of C_{gs} were between the gate and the channel region, our value for C in Eq. (4.15) would itself be a variable and any calculation would be more difficult. As it is, we can place a value for threshold voltage into ∂V and the time we wish our switch to operate as ∂t and have a reasonable idea of the driving current required to charge the capacitor in that amount of time.

But that is only the start of our problem. Once we reach and pass the threshold level of our power MOSFET, things begin to happen. Drain current begins and our power FET's forward transconductance begins to build up. In our discussion on capacity we analyzed the Miller effect and its impact on input capacity. In Eq. (4.9) we saw the effect that voltage gain has on input capacity. Earlier in this chapter we determined that the forward transconductance of these short-channel power MOSFETs saturates at a predetermined gate voltage V_G [see Eq. (4.11)]. Putting all this together, we can see that as the gate voltage V_G rises, both the drain current and the forward transconductance rise and eventually saturate. However, voltage gain A_V is directly proportional to transconductance, so we also see voltage gain rising and eventually saturating. If the voltage gain is, therefore, proportional to gate voltage V_G, we can expect that the Miller capacitance will also be proportional. What all this means to us is that as we continue to increase the gate voltage to turn the MOSFET more fully ON, our input capacity rises quickly to possibly enormous proportions, depending upon the feedback capacity C_{gd}.

If our input capacity is now variable and rising swiftly, our equation for determining the drive current becomes complicated. We would do best if we simply calculated the worst-case input capacity and made our calculations for worst-case drive current.

Of course, once we reach a full ON condition, we no longer need the current simply because the expression in Eq. (4.15), ∂V, is zero. We now have a steady-state situation.

Although we may have properly calculated the necessary drive current needed to switch our power MOSFET ON, the question that may arise is: Do we have a source sufficient to deliver the necessary current in the required time? The rise time of our power MOSFET depends upon the total series gate resistance, which, among other things, includes the source resistance of the driver stage, and, of course, the total effective input capacity of the power FET. This rise time depends upon how many *time constants* have elapsed from our initiation of current. Using the language of analog switching, we consider that the input waveform must rise from the 10% to the 90% level in a prescribed time interval. This time interval equation is well known:

$$t_r = 2.2RC \qquad\qquad (4.16)$$

where R is the total series gate resistance, including the driver output impedance, and C is the total effective input capacity of our power FET, including the Miller effect.

Our power FET is ON and we now want to switch it OFF. It would seem reasonable that we need only remove the gate voltage from our power MOSFET and we would accomplish our goal. Not quite. Our input capacitor stands at full charge, and before we can remove the gate voltage we must concurrently withdraw the charge. Whatever the total resistance was during the charge now become the discharge path for the charge current. The power FETs do not have minority carriers, so once we have successfully discharged our input capacitor, we have also successfully turned OFF our MOSFET switch.

We should take special notice of the *total* series gate resistance that affects our switching speed. Many power MOSFETs are constructed with polysilicon gates, whereas others have metal gates. To gain a better appreciation of the effect that a polysilicon gate might have on switching time, consider that the sheet resistance can be as high as 200 Ω/square, whereas for a metal gate the sheet resistance is practically nil. Furthermore, when a polysilicon gate structure is used, we find a type of transmission line effect consisting of series resistance and distributed capacity as shown in Fig. 4-26. Because of the time delay

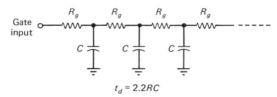

$$t_d = 2.2RC$$

Figure 4.26 Equivalent circuit of a polysilicon gate, showing the distributed capacity effects.

associated with such a transmission line, a large-geometry power MOS-FET using a polysilicon gate structure might be severely limited in turn-ON time merely due to the time constant alone. A metal gate does not suffer so badly and would therefore find more popularity in those applications requiring very fast switching or in high-frequency applications.

It should be apparent to us that switching a power FET is a study in charge transfer. Because of the nearly-infinite dc input resistance of the power FET, about all that is necessary is to know how to charge and discharge the equivalent input capacitance, remembering, of course, that there are *three* states involved in the "turn-ON" cycle: the prethreshold state, the initial turn-ON state, and the full-ON state. In the prethreshold state, shown in Fig. 4–27, the input capacity is, for the

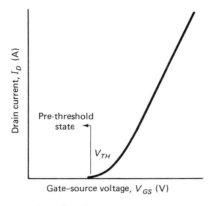

Figure 4.27 Typical transfer characteristic curve, showing the prethreshold, or OFF state.

most part, simply the gate capacity C_{gs}, and our charge transfer can be readily calculated using Eq. (4.17).

$$W = \tfrac{1}{2}C\partial V_G{}^2 \qquad \text{watt-seconds} \qquad (4.17)$$

During the initial turn-ON state, the Miller effect enters the equation and the effective input capacity becomes rather complex. We need to be aware that during turn ON several interrelated events begin that complicate our defining of the input capacity. We can use Fig. 4–28 to follow the sequence. As the drain current ramps upward, so does the transconductance and with it the voltage gain A_V. As the voltage gain increases, the Miller effect becomes more pronounced [see Eq. (4.9)]. But as the drain current increases, we also see the voltage drop across the power FET drop, simply because of the IR drop across the load resistor. In Fig. 4–14 we saw how the feedback capacity changes with V_{DS}, so we can reasonably anticipate that as the drain current ramps upward, C_{gd} will also change.

When we reach the full ON state the Miller effect no longer predominates but because of channel current we can now observe a somewhat higher value of overall gate capacity than we did during the prethreshold state because of the increase in gate-to-channel capacity C_{gc}. When we turn the power FET OFF, we also enter into several identifiable states, but now our major concern is only how to sink the charge current quickly to turn the FET OFF.

It is easy for us to envision two rather simplistic models of the

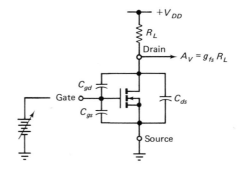

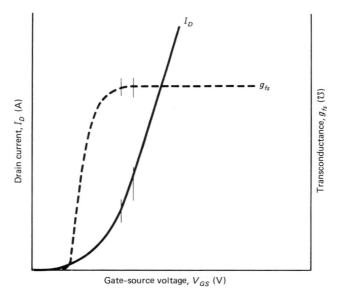

Figure 4.28 Superimposed pair of curves, showing the relationship between the static transfer characteristic and the forward transconductance of a power MOSFET.

power FET as a switch using Fig. 4-29. In its simplest form the ON condition consists of $R_{DS(on)}$ shunted by our parasitic p-n diode. In the OFF state we simply have the p-n diode as a series element. Because of this series diode it becomes rather obvious that we cannot use the vertical power MOSFET as an analog switch without special circuit design, which we discuss in Chap. 9.

High-speed switching and easy interfacing between the power FET and logic are certainly two important features that have made the

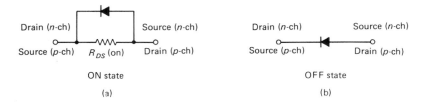

Figure 4.29 Basic switch models of the power MOSFET, showing in (a) the ON state, and in (b) the OFF state.

TABLE 4-1 Comparison of Three Logic-Driven Devices

Part number	Chip area (mils²)	Breakdown voltage (V)	Max $R_{DS(on)}$ or $V_{CE(sat)}$ at I_C/I_B (A)	I_L (A)	Output loss (W)	Drive loss (W)	Total loss (W)
2N5758/60	24.3 k	100/140	1.0 V at 3/0.3	3	3.0	3.6	6.6
MOSFET	22.5 k	100/120	0.1 Ω	3	0.9	- -	0.9
2N6052	24.3 k	100	2.0 V at 9/0.024	6	12	0.3	12.3
MOSFET	22.5 k	100	0.1 Ω	6	3.6	- -	3.6
2N6545	24.3 k	400	1.5 V at 5/1	5	7.5	12	19.5
MOSFET	22.5 k	400	1.0 Ω	5	25	- -	25.0
2N5629/31	38.6 k	100/140	1.0 V at 10/1	10	10	12	22.0
2N6384	40.0 k	100	2.0 V at 10/0.04	10	20	0.5	20.5
MOSFET	40.0 k	100/120	0.05 Ω	10	5	- -	5.0
2N6547	40.8 k	400	1.5 V at 10/2	10	15	24	39.0
MOSFET	40.0 k	400	0.5 Ω	10	50	- -	50.0

Drive losses computed from circuits shown below

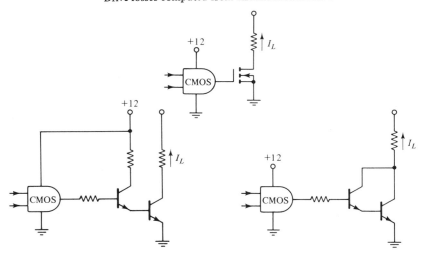

power MOSFET so popular. There is, however, another important feature that often goes unrecognized but not unnoticed, and that is the power loss involved in the switching cycle. In a switching cycle we find three types of losses: drive losses, output losses, and switching losses. An interesting comparison is offered in Table 4–1 among a logic-driven power FET, a logic-driven bipolar transistor, and a logic-driven Darlington pair. The comparison is made between power transistors of comparable size and capacity. We should be careful to observe that since most power FETs can be driven directly from logic, we omitted the driver and as a consequence, show no drive loss. Output loss is simply the power dissipated across either the bipolar transistor, $I_C V_{CE(sat)}$, or the power MOSFET, $I_D^2 R_{DS(on)}$.

A major concern involves the switching loss we might experience, and this loss depends to some extent upon the load. We would expect that the current–voltage relationship passing through our power FET switch would be different for resistive and reactive loads. We have already discussed switching loss for resistive loads, for such loss is simply the $I^2 R_{DS(on)}$ loss, which we show in Table 4–1. Reactive loss involves the power-factor contribution by the load; that is, does current lead or lag the voltage and by how much? Figure 4–30 provides us with a clear understanding of what we can expect for a highly inductive load during switching.

There is much that we can say with respect to switching that perhaps lies outside the scope of this book. The reader is invited to

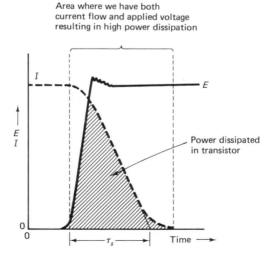

Figure 4.30 Power-factor contribution to internal dissipation during switching.

research the extensive literature offered in the References concluding the chapter.

4.5 MODELING FOR AMPLIFIERS

When trying to develop an electrical model of a transistor, it is often helpful first to visualize the physical structure. Then from this, and with an understanding of its operation, we can literally trace the major electrical circuit elements and develop a reasonable model. There are, to be sure, some circuit elements that cannot be physically visualized— such as the current generator $v_{gs}g_m$ and the output conductance g_{os}— but these we understand and so we can include them in our model.

Our first model begins with the cross-sectional view of the static induction transistor (SIT) in Figs. 2-16 and 3-2. The depletion area about the gate acts as plates of a capacitor coupling both to the drain and source. Finite resistance appears in series with both capacitors as a result of the p-doped gridded gate structure as well as the epitaxy and the depletion region. With some careful study we can soon evolve the electrical circuit shown in Fig. 4-31. As a matter of fact, this model bears a close resemblance to the common J-FET, and so it should, because it is a J-FET. What is noticeably absent is the drain-to-source capacitance C_{ds}, simply because the gate structure acts as a rather effective shield, and the characteristic source resistance R_s. When we were introduced to the SIT back in Chap. 2 we read where the inventors had developed this special type structure principally to reduce or hopefully eliminate the parasitic channel resistance. We know, of course, that they were successful because we see an unsaturated drain current with increasing drain voltage. This extraordinarily low source resistance and the melding of the output capacity into what appears as feedback capacity, C_{gd}, are the two major variants between the SIT and the more common planar J-FET.

The low-frequency version of the SIT is generally used in the *common-source* configuration, where the source is common to both the input and output signals. Because of the low source resistance we might expect a high-voltage gain, as degeneration is, for the most part, absent. Unfortunately, we are dealing with a solid-state triode whose output conductance is quite high. As a consequence, our voltage amplification, more commonly called *mu*, μ, is about equal to what we would expect from a power vacuum-tube triode. Mu, or voltage amplification, may be calculated from the equation

$$\mu = \frac{g_m}{g_{os}} \qquad (4.18)$$

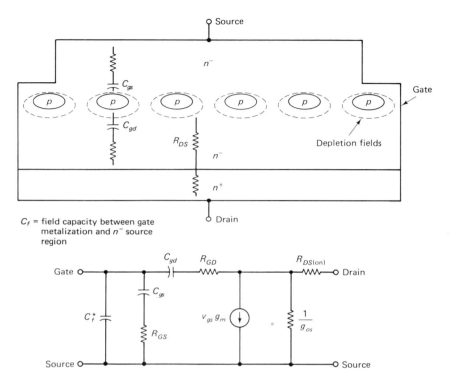

Figure 4.31 Basic common-source amplifier model of the static induction transistor.

Most power SITs offer a voltage-amplification figure of somewhat less than 10.

The high-frequency SIT, especially fabricated to reduce as much of the parasitic input capacitance as possible, often is used in the *common-gate* mode. With the gate at RF ground, the feedback capacity from drain to gate C_{gd} is effectively muted insofar as raising the possibility of unwanted feedback or Miller effect. But again, the fundamental problem lies with mu, or low-voltage gain. However, the SIT does offer potential for a power amplifier for single sideband, or linear service in high-frequency systems because of its desirable triode output characteristics. The grounded gate model would look like that in Fig. 4–32. Again we need not be reminded of the similarity with the planar J-FET, for in the common-gate mode there is no observable difference.

Modeling the power MOSFETs can be done in much the same way. If we start with a physical model, that is, an accurate cross-sectional diagram such as we saw in Figs. 2–28 and 2–29 (for the VDMOS and VMOS, respectively), we can do a remarkable job in arriving at an electrical equivalent. Generally, the problem we face is

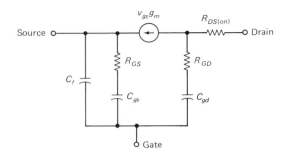

Figure 4.32 Basic common-gate amplifier model of the static induction transistor.

that our model is too exact and unduly complicated for simple circuit analysis.

When we take the task to hand and begin modeling our electrical circuits from the cross-sectional view, we must remember that we have a parasitic *npn* bipolar transistor intrinsic to the DMOS and VMOS power MOSFETs. If we fail to include its effects, we will end up where most others have tried to model these MOSFETs. What they witness is a measured power gain versus frequency that progressively deteriorates with increasing frequency from what they have modeled. The typical model, excluding the *npn* bipolar transistor, generally looks like the circuit in Fig. 4–33. It takes little searching through the literature to find an equivalent circuit model that looks remarkably similar to this. Generally, they do not work, at least not at high frequencies. For low-frequency and dc applications they are perfectly good and useful because they are not complicated.

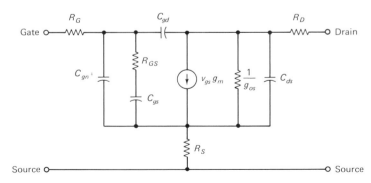

Figure 4.33 Typical electrical model of a short-channel power MOSFET, excluding parasitic elements.

We will not dwell on the equivalent circuit shown in Fig. 4–34, which is what we might call an *exact* equivalent circuit; if you carefully

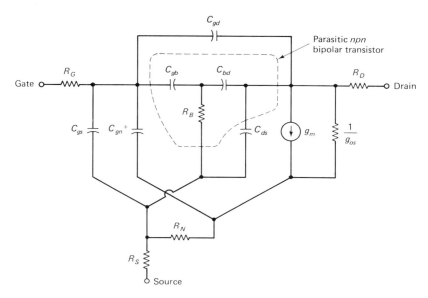

Figure 4.34 Exact electrical model of a short-channel vertical MOS-FET. (Courtesy *RF DESIGN* © 1979 Cardiff Publishers.)

follow either Fig. 2–28 or Fig. 2–29, you will get the same circuit. Although perhaps *exact*, it is messy to analyze, which is another reason why we generally resort to simple models, as we saw in Fig. 4–33. We should take care to note the electrical contribution of the parasitic *npn* bipolar transistor (in the dashed area). Since we feel reasonably confident that we can prevent the base of this parasitic from going positive, we only need to model its parasitic *elements*, which form a feedback network. No wonder that when one fails to include this parasitic, the measured gain falls off. We have a parasitic feedback path from drain to gate, from output to input. An interesting observation is to compare the calculated effect of including the parasitic with the calculated gain when the parasitic is not included. We see the results in Fig. 4–35.

Our simplest electrical model for either MOSFET is the common-source circuit shown in Fig. 4–33. The current generator is simply $g_m v_{gs}$ and the voltage gain across the load would be simply

$$\frac{e_{\text{out}}}{e_{\text{in}}} = \frac{-gmR_L}{1 + R_L g_{\text{os}}} \tag{4.19}$$

For our power MOSFETs g_{os}, the output conductance, is quite low and our voltage gain is quite good. Output conductance is simply the ratio $\Delta I_D / \Delta V_{DS}$. A MOSFET with a "pentode" output characteristic will by nature have a low value for g_{os}, which is very desirable.

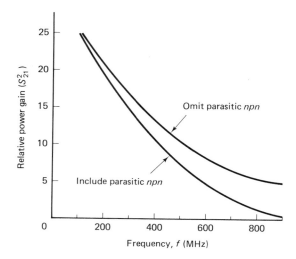

Figure 4.35 Relative effects of including or excluding the parasitic *npn* bipolar transistor elements in the calculation of power gain. (Courtesy *RF DESIGN* © 1979 Cardiff Publishers.)

A common-gate model serves no useful purpose. Power MOS-FETs, unlike planar J-FETs (which we often find in common-gate circuits), have a unique structure that forbids us from reducing the feedback capacity if we ground the MOSFET's gate. In the case involving the J-FET, grounding the gate effectively places a Faraday shield between drain and source. Not so with our vertical power MOS-FETs—one glance at their cross-sectional views (Figs. 2–28 and 2–29) shows that this will not happen. Using a vertical power MOSFET in the common-gate mode actually increases the feedback capacity from what was C_{gd} to C_{ds}.

Summing up this section we find that the SIT very closely follows the J-FET model with the notable exceptions of lower source resistance and lower output capacity. For the high-frequency model the input capacity is also reduced. The vertical power MOSFETs model identically (aside from actual numerical dimensions assigned to the parameters) and we have a variety of simple models from which to choose. The exact model fits best for high-frequency applications, whereas the model that conveniently omits the parasitic works well at midfrequencies (below 10 MHz) and the very simple common-source model is ideal for low-frequency and dc applications. If we are to consider using *computer-aided design* (CAD) techniques, we might find the exact model most useful regardless of whether we plan a low-frequency or midfrequency design.

Now that we have a handle on our models, we should be able to gain a better insight into the problems that beset us when we use power FETs. At first we might assume that a power FET would be the ideal power transistor and that the comparisons made in Chap. 1 were very convincing. Because of the power FET's thermal degeneracy we should not have the problems so familiar with power bipolar transistors, and without minority-carrier storage time we should not find problems in switching. Are we now going to admit that there is a problem with our power FETs? If we need to answer before we continue, we will have to reply with both a qualified *yes* and a qualified *no*, which really does not answer our question at all. One question that presses for an immediate answer is: What does "SOA" mean—what is the *safe operating area* of a transistor?

Whenever we prepare to use a power transistor we always go to the data sheet looking for the operating limits on performance. Universally, these limits are breakdown voltage, maximum permissible operating current, and thermal dissipation. We can plot these limits, preferably on log-log paper, and by doing so we can claim that we have the safe operating limits for our transistor. The enclosed *area* becomes the safe operating area (SOA). If we turn back to Fig. 1-7, we can see an idealized SOA for a power transistor. With this graph before us we should feel confident that our power transistor will work safely anywhere within the boundaries of the SOA. Right?—It is still too soon to answer.

Let us first discuss the safe operating area of a "typical" power bipolar transistor. Our boundaries are: for the ordinate, the collector current I_C; and for the abscissa, the collector-to-emitter voltage V_{CEO}. The thermal dissipation limit closes our area. Unfortunately, for the greater majority of power bipolar transistors our SOA is not yet enclosed. What is lacked is the effect of a phenomenon called *second breakdown*, which once added to our graph closes the area of our SOA plot. Once we have added its contribution we discover that our safe operating area is somewhat restricted. This we can see if we return to Fig. 1-7. At *primary* breakdown BV_{CEO}, we see a sudden surge of collector current but not to the limit of thermal dissipation. Instead, we discover what looks like a negative slope, further reducing our maximum operating voltage while increasing our collector current. This strongly suggests (and, indeed, *very truthfully*) that if we operate our transistor near breakdown and for even a moment pull sufficient collector current, we might find ourselves—and our power bipolar transistor—operating *outside* the SOA.   In Chap. 1 we learned that power MOSFETs *can* be forgiving when accidentally stressed beyond their SOA, but bipolar transistors are *never* forgiving. They burn out.

There have been numerous theories to support our understanding of second breakdown, but the one that appears to come closest contends that an uneven distribution of current beneath the bipolar emitter develops an electric field that far exceeds the critical value necessary for ionization. The result is a premature avalanche breakdown. The final result, or effect, is a catastrophic rise in junction temperature that can literally either melt or fuse the silicon, which destroys the transistor. We can anticipate that short pulses of V_{CEO} and I_C will allow us to "push out" the SOA to more nearly identify with the ideal situation.

Now let us discuss the power MOSFET. The fundamental cause of second breakdown is very similar to that offered before for the power bipolar transistor, that is, the development of an electric field that far exceeds the ionization level of the material followed by avalanche breakdown. Among the References found at the end of the chapter are many highly technical papers that have tried to delineate the sources of second breakdown in power transistors, including, of course, power MOSFETs. We may perhaps oversimplify one principal cause for second breakdown in the following discussion for the sake of clarity.

Earlier, in Sec. 4.3.2, we developed a model of the power MOSFET that included a parasitic *npn* bipolar transistor. We may remember from Chap. 2 that we were explicit as to the importance of keeping this parasitic transistor OFF.   We even discussed the consequences that would result if the parasitic transistor were ever inadvertently turned ON.   Second breakdown results.   Now we are ready for a question: How do we "inadvertently" turn ON this parasitic *npn* bipolar transistor, and how does that cause second breakdown?

To begin, we address the first part of this question.   Again, if we reflect back to Chap. 2, we recall that power MOSFETs are fabricated with an electrical bridge spanning from source to base. That, we were told, was to tie the parasitic bipolar transistor's base to its emitter. That *should* keep the *npn* transistor OFF. But does it?   We also mentioned briefly that the *p*-doped channel has a finite resistivity and unless care was taken in the initial design we could induce a charge that would effectively bias this *npn* bipolar transistor ON, even if the base *appeared* shorted to the emitter. This might occur with the injection of carriers into the *p*-doped channel since the parasitic bipolar transistor *appears* close to the inversion layer of the MOSFET. If our power MOSFET is stressed to the point where it is carrying a high current and withholding a high voltage, then it is entirely possible, and probable, that we have excited an intense electric field containing highly charged electrons. Investigators have determined that these highly charged electrons are in the n^- drain–drift region. Because of the high voltage we have impressed across the MOSFET, these electrons are accelerated toward the source. Another look at the cross section of a typical power MOSFET (Fig. 4-20) will reveal an obvious barrier: the p-doped channel. For both our VMOS as well as our VDMOS, this channel is the target for these highly charged electrons. Unless we have taken special precautions in our MOSFET design we may activate the parasitic npn bipolar transistor and open the door to second breakdown.

Now we can turn our attention to answering the second part of our question. Once we activate this parasitic bipolar transistor, which is shunted across the power MOSFET and thereby must withstand the entire voltage, its breakdown BV_{CER} is appreciably less than its BV_{CBO} and less than the power MOSFET's BV_{DSS} (which, incidentally is equal to the bipolar transistor's BV_{CBO}). Since in our original qualifying statement we said that we were stressing the power MOSFET both with current and high voltage, we have, in fact, exceeded the breakdown of our parasitic bipolar transistor and avalanche breakdown—second breakdown—has occurred.

Before we conclude both this section and this chapter, we must consider the mechanics that might precipitate second breakdown in our power MOSFETs. A pure resistive load is definitely safe, but a reactive load may lead to trouble. Our problem is power factor; that is, as we discussed in Sec. 4.4, the current may either lead or lag the voltage, resulting in a situation where we have both high current *and* high voltage simultaneously. It is precisely this that brings about intense electric fields, which, in turn, can precipitate second breakdown. What we saw in Fig. 4-30 describes the problem perfectly. The first prevention we may use is to either raise the breakdown rating of our power MOSFET used in this reactive load application or, for an inductive load, place a free-wheeling diode across the inductive load or even a zener across the MOSFET. The circuits that we study in later chapters have been designed to reduce or eliminate the problems of second breakdown.

REFERENCES

BEATTY, BRENT A., SURINDER KRISHNA, and MICHAEL S. ADLER, "Second Breakdown in Power Transistors Due to Avalanche Injection," *IEEE Trans. Electron Devices*, 23 (1976), 851–57.

DECLERCQ, MICHAEL J., and JAMES D. PLUMMER, "Avalanche Breakdown in High-Voltage D-MOS Devices," *IEEE Trans. Electron Devices*, 23 (1976), 1–4.

GHANDHI, SORAB K., *Semiconductor Power Devices*. New York: John Wiley & Sons, Inc., 1977.

HERSKOWITZ, GERALD J., and RONALD B. SCHILLING, eds., *Semiconductor Device Modeling for Computer-Aided Design*. New York: McGraw-Hill Book Company, 1972.

HU, CHENMING, "A Parametric Study of Power MOSFETS," *Conf. Record*, Power Electronics Specialists Conference, 1979, Paper 5.7.

KAY, STEEVE, C. T. TRIEU, and BING H. YEH, "A New VMOS Power FET," *Technical Digest*, IEEE Electron Devices International Meeting, Washington, D.C., 1979, pp. 97–101.

KRISHNA, SURINDER, "Second Breakdown in High Voltage MOS Transistors," *Solid-State Electronics*, 22 (1977), 875–78.

LAZARUS, M. J., "The Short Channel IGFET," *IEEE Trans. Electron Devices*, 22 (1975), 351.

LIDOW, A., T. HERMAN, and H. W. COLLINS, "Power MOSFET Technology," *Technical Digest*, IEEE Electron Devices International Meeting, Washington, D.C., 1979, pp. 79–83.

National CSS, Inc., *Ispice Short Channel MOSFET Model Note*, Norwalk, Conn., 1977.

OMURA, Y., and K. OHWADA, "Threshold Voltage Theory for a Short-Channel MOSFET Using a Surface Potential Distribution Model." *Solid-State Electronics*, 22 (1979), 1045–51.

POCHA, MICHAEL D., and ROBERT W. DUTTON, "A Computer-Aided Design Model for High Voltage Double Diffused MOS (DMOS) Transistors," *IEEE Jour. Solid-State Circuits*, 11 (1976), 718–26.

POCHA, MICHAEL D., J. D. PLUMMER, and J. D. MEINDL, "Tradeoff between Threshold Voltage and Breakdown in High Voltage Double Diffused MOS Transistors," *IEEE Trans. Electron Devices*, 25 (1978), 1325–27.

RICHMAN, PAUL, *MOS Field-Effect Transistors and Integrated Circuits*. New York: John Wiley & Sons, Inc., 1973.

TEMPLE, V. A. K., and P. V. GRAY, "Theoretical Comparison of DMOS and VMOS Structures for Voltage and On-Resistance," *Technical Digest*, IEEE Electron Devices International Meeting, Washington, D.C., 1979, pp. 88–92.

TOYABLE, TORU, KEN YAMAGUCHI, and SHOJIRO ASAI, "A Numerical Model of Avalanche Breakdown in MOSFETs." *IEEE Trans. Electron Devices*, 25 (1978), 825–32.

WANG, PAUL P., "Device Characteristics of Short-Channel and Narrow-Width MOS-FETs," *IEEE Trans. Electron Devices*, 25 (1978), 779–86.

YOSHIDA, ISAO., et al., "Thermal Stability and Secondary Breakdown in Planar Power MOSFET's." *IEEE Trans. Electron Devices*, 27 (1980), 395–98.

five

Using the Power FET in Switching Power Supplies and Regulators

5.1 INTRODUCTION

The demand for efficient energy conversion is not restricted solely to the automobile. We are in a new world where energy costs are a major expense. We have been aware of the overall inefficiency of linear inverters, regulators, and power supplies. None of them capable of delivering power of any consequence has ever run cool. Energy losses through thermal dissipation is well known, and equally well known is the improved efficiency possible in switching systems. We have witnessed tremendous growth in the switching power supply market and daily we find new equipment replacing the old linear power supply with lightweight, cool-operating switching power supplies, commonly called *switchers*. Switching involves little power loss if the switching element is efficient. In an earlier generation auto radios used a primitive switching power supply to convert 6 V to several hundred needed to power the vacuum tubes. Here they used synchronous mechanical vibrators, half of the unit breaking up the 6 V into ac pulses feeding the power transformer, the other half of the synchronous vibrator to "rectify" the high voltage. Old timers will vouch that these units ran a whole lot cooler than rectifiers! But, unfortunately, they were not efficient. Efficiency does not stop at the rectifier; we need filtering, and these old synchronous vibrators were pretty low frequency devices and required massive filters. If the bipolar transistor and the 12-V battery had not entered the market, perhaps we would have seen the evolution of switching power supplies a generation earlier. Not until the power bipolar transistors and the high-power SCRs arrived did we enjoy efficient energy conversion. But even with these solid-state devices we knew that efficiencies could be improved.

Before we proceed further we must define *efficiency* as we plan to use it in this and subsequent chapters. Efficiency must be more than simply the ratio of power out to power in; for many it entails the efficient utilization of funds, in other words, how much it costs. Reliability is part of that package. An unreliable product, one whose mean-time-between-failure (MTBF) is known to be short, will require us to keep a spare "on the shelf." That is inefficient. A major cost savings is, of course, to reduce the number of components and then to reduce the cost of the components we use. As we reduce the number of components, we also improve the MTBF provided that we maintain reasonable operating temperatures and do not stress the components unduly by working them either at or too close to their ratings. The U.S. government's MIL-HDBK-217B (*Military Handbook*) offers recommendations for reliability based on thousands of hours of study that we would be wise to consider.

Based on the conclusions that we have drawn from previous chapters, we can build a strong argument to justify our need for power FETs in the design of high-efficiency switch-mode energy converters. For example, let us assume that our switch-mode converter is to be designed around a logic element, perhaps a microprocessor or even one of the popular switching regulator controllers such as the TL494. We know that it is easier to drive a voltage-controlled power FET than it is a current-controlled power bipolar transistor, so immediately we can envision fewer intermediary parts between our logic element and the power transistor, in this case a power FET. Another illustration might focus on cost reduction of the parts we do use. Transformers take a very prominent role in energy conversion and their cost is certainly proportional to their size. Small transformers generally cost less. Large bulky transformers are often identified with power, but we should remember that we usually associate such transformers with linear energy converters operating at 60 Hz. For more reasons than cost savings on the transformers, switch-mode converters operate between 20 and 60 kHz. We discuss why later in this chapter.

We can easily accept that a transformer's size will shrink as we increase its operating frequency, however, the cost reduction between, say a 60-Hz power transformer and one designed for operation at 20 kHz is not dramatic. The major improvement we discover is the degree of filtering as compared to that required for a linear, 60-Hz energy converter.

So far we have focused our attention on what many of us might agree is obvious; that is, a switch-mode energy converter is more efficient than a linear converter, and the higher the operating frequency, the more efficient it becomes as to weight, size, and possibly cost. Perhaps not immediately obvious to us is that the higher the speed (frequency), the shorter the recovery time. Since the purpose of this chapter is to justify using power FETs in such applications, let us again turn our attention to a review of the features that power FETs offer and the benefits that we can gain by using them in our designs.

5.2 POWER FETS VERSUS BIPOLAR TRANSISTORS

Higher speed is higher frequency. Many switch-mode energy converters using power bipolar transistors are limited to frequencies ranging from 20 kHz to something under 100 kHz, depending on their delivered power. Some switchers operate even higher than 100 kHz but for low-power applications. The active element, whether we are using a bipolar transistor or a power FET, must be capable of maintaining peak performance while operating at high currents and voltages and into highly reactive loads (mainly inductive).

In Chap. 1 we identified two advantages that power FETs have over power bipolar transistors. We first saw that the power FET as a bulk semiconductor operates exclusively as a majority-carrier transistor and as a consequence has no minority-carrier storage time. Second, as a bulk semiconductor, the power FET has degenerate thermal properties. These two properties suggest to us that we should be able to switch a power FET very fast, and if we inadvertently overstressed the FET, there would be a good chance of recovery.

It is true, of course, that power bipolar transistors can be made to switch very fast—not as fast as a comparably rated power FET— but, nonetheless, at very respectable speeds. The only problem is that to do so requires special driving techniques to ensure that we do not push them into saturation, where storage time becomes excruciatingly long. A power bipolar transistor operated out of saturation will show a greatly reduced storage time. Drive circuitry to accomplish this improvement in switching speed can take numerous forms. Two popular methods that we can use to shorten the storage time of bipolar transistors are evident in Fig. 5-1. By far the simplest method is to use a capacitor in parallel with a resistor. When we drive the bipolar ON the voltage drop across the resistor charges the capacitor. When we turn the bipolar OFF, for example, by reverse biasing the

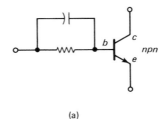

(a)

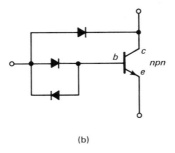

(b)

Figure 5-1 Two popular circuits used to shorten the switching time of bipolar transistors. (a) RC input speeds switching-OFF time. (b) The Baker clamp prevents the collector voltage from falling below the base voltage.

base, the charge on the capacitor sweeps the minority carriers out of the base and thus effectively overcomes the storage time delay. Another method is using a pair of diodes to prevent the collector from becoming forward-biased. This scheme is called the Baker clamp.

There are, or course, other ways by which we can reduce the storage time of a power bipolar transistor. For example, we can modify the drive circuitry to provide a substantial *negative* pulse that, like the capacitor (Fig. 5–1), sweeps the minority carriers out of the base. Still other methods include such drives as the current mode of operation. What we are witnessing is the obvious complication required of driving circuits to operate power bipolars at high switching speeds. It is pretty obvious to us that base drive is critical to efficient high-speed switching with bipolar transistors. Even with a successful base drive we are bound to experience transition times—due to minority-carrier storage— of at least an order of 10 times what we would be able to measure with power FETs! We also recognize that we should use bipolar transistors with the highest f_t available. Unfortunately, as their f_t increases, their BV_{CEO} decreases—a characteristic that we certainly do not want for a switcher.

A power transistor used in a switcher really takes punishment and for the sake of reliability has to be underrated. The MIL-HDBK-217B handbook suggests that we keep under 80% of full-rated current and voltage, which, at best would be difficult for many of us. Thermal stress is undoubtedly the major cause of power bipolar transistor failure in the field. This failure mechanism reveals itself to us as second breakdown.

Power FETs, especially power MOSFETs, simply do not exhibit these problems. We have little trouble driving them and certainly no trouble achieving fast switching. We can, if we need, raise their switching speed by at least an order of magnitude beyond what would be either possible or practical for a comparably rated power bipolar transistor. If, in our use of power FETs, we discovered that our drive power was insufficient to charge the input capacitance of the power FET fully in the time desired, we can easily implement complementary emitter followers as predrivers, as shown in Fig. 5-2. When we ramp up an input voltage to their bases, the *npn* effectively ties the MOSFET's gate to the positive rail. Conversely, when we ramp down, the *pnp* clamps the gate to the ground. In either case the charging and discharging time constant is greatly reduced.

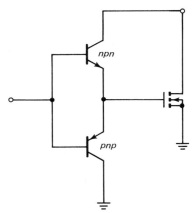

Figure 5-2 Emitter-coupled drive lowers the drive impedance and greatly speeds up the switching time.

Another significant feature of using voltage-driven power FETs becomes apparent when we build either a half-bridge or full-bridge switcher, which necessitates us stacking the power FETs in what is commonly called a totem pole. A comparison of totem pole configurations between bipolar transistors and power MOSFETs is provided in Fig. 5-3. We can trace the operation of the totem pole, beginning with the MOSFETs, by noting that with a positive-going input, FET a and FET c turn ON. FET a pulls the gate of FET b down, ensuring that FET b remains OFF. When our input signal drops both FETs a and c turn OFF, and the gate potential on FET b rises to the positive rail, ensuring that FET b turns ON hard. The cycle can then be repeated. The bipolar totem pole operates similarly. A positive pulse turns ON BJT a and b; a pulls down the base of BJT c, ensuring a continual OFF condition. A negative-going signal on the bases of a and b turn both

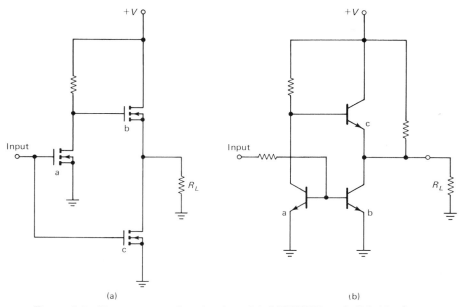

Figure 5-3 Two totem-pole circuits: (a) MOSFET and (b) bipolar transistor.

OFF and c turns ON. We can imagine the problems we would face if we were trying to switch totem-pole bipolar transistors fast.

Since the secret to high-speed switching is the drive, we can appreciate why the power FET should be our first choice. If we tried to speed up the switching speed of a bipolar transistor totem-pole arrangement and for any reason inadvertently got both bipolar transistors, b and c (in Fig. 5–3), ON, we would have a short circuit.

Another very simple and effective means for us to drive a power MOSFET totem pole, which has special application in switchers, is to use an isolation transformer for the uppermost MOSFET of the pair. Since we would be running our system at a fairly high switching frequency (at least 100 kHz), our isolation transformer would be a small pulse transformer, which, as shown in Fig. 5–4, would be driven by complementary emitter-coupled bipolar transistors and which, in turn, would drive the MOSFET gate. The secondary would be, in effect, floating with respect to ground. The speed of such a totem-pole driver could be extremely high, an impossibility had we chosen to use bipolar transistors in a half-bridge circuit.

While on the subject of totem poles we should address the full-bridge circuit, that is, where, like the figure H, we have power MOSFETs in each leg. To effect alternating current flow through the cross-arm where we would place our main transformer, we must *alternately*

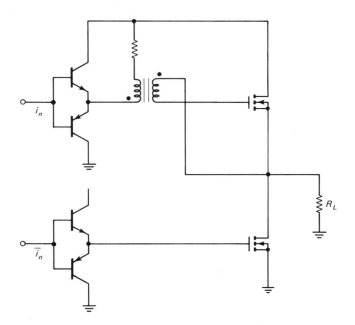

Figure 5-4   Totem-pole power MOSFETs using a pulse transformer to couple to the upper MOSFET.

switch the power MOSFETs in a sequence of "one high, one low," and "one low, one high." Because of the relatively easy drive requirements of power MOSFETs, we can easily implement this, as we show in Fig. 5-5. The pulse transformers for this arrangement can be very simply constructed using trifilar wire wound on ferrite bobbins. With but a little care we should easily be able to switch at frequencies beyond 1 MHz!

An early advantage that bipolar transistors enjoyed was their superior current-handling ability based on the simple fact that we could obtain larger devices than we could power FETs. However, this early advantage has failed to keep pace with the technology. Today if we found that we needed higher current capacity than what we were able to handle with our present power MOSFETs, paralleling is simplicity itself. Not so with bipolar transistors. To ensure that our parallel power MOSFETs share equally, we must watch two things. First, we should select devices with a reasonably close match in threshold voltage V_{TH}, and second, have both devices on the same heat sink. We must remember that not only does the current handling double for a pair, but also the input capacity, so we must have the capacity to drive them if speed is important to us. If our decision is to parallel a number of power MOSFETs we would do well to string a small ferrite bead onto each gate lead to inhibit any possible parasitic oscillation. If ferrite

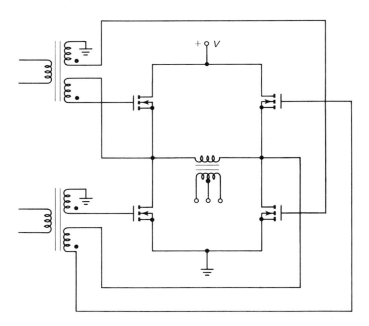

Figure 5-5   Full-bridge totem pole driven by pulse transformers.

beads are unavailable, a series gate resistor in the neighborhood of 30 to 50 Ω is advisable, as shown in Fig. 5-6.

If we were to compare the switching losses of a bipolar transistor with those of a power MOSFET, we would discover that as we raised the switching speed we would see a definite improvement in the performance of our power MOSFET and a gradual degradation in the performance of the power bipolar transistor. At low switching speeds, where we find most bipolar transistors operating today, we might be hard-pressed to substitute a power MOSFET and obtain equal or better

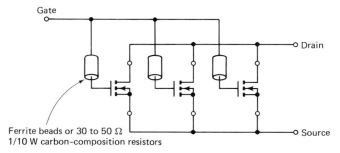

Figure 5-6   Paralleling power MOSFETs to prevent unwanted parasitic oscillations can be done by stringing ferrite beads on each gate wire or, as an alternative, inserting low-value carbon resistors in series with each gate lead.

efficiency. Of course, our intention is not simply to substitute power MOSFETs for bipolar transistors but to improve reliability, reduce the number of components, raise the efficiency, and hopefully, trim the cost. From what we have already covered, we are developing an intuitive appreciation for a need to raise the switching frequency, but not by merely a decade or so, but by at least an order of magnitude. If our bipolar transistor switching energy converter had been operating at 35 kHz, we would want to push the speed to at least 350 kHz, or more.

5.3 POWER FETS VERSUS THYRISTORS

In Chap. 1 we briefly identified one major shortcoming of the thyristor, or SCR, as having a slow turn-OFF time. There are, indeed, other shortcomings which we should examine, but before we do we should at once set aside the notion that the power FET is now ready to do battle with the SCR. If we were to stumble across a moderate-size switcher where SCRs were used, we might have a case to argue, and that we do in this chapter. However we should be painfully aware that the SCR has practically no limit to power-handling capacity. Major power-generating stations use SCRs to control megawatts of electric power. Welders use SCRs to control hundreds of amperes of current. Induction heating equipment has used SCRs for years in steel mills where hundreds of megawatts of power are controlled. We might well wonder if our power FET will ever replace such energy converters.

The slow turn-OFF time attributed to the SCR results from the fact that once the SCR has been triggered, commutation of the anode voltage is required to shut it down. Like the bipolar, it then needs a finite time for the minority carriers to bleed away or to recombine. Other problems plague the SCR, especially if high switching speeds are contemplated. One problem is that before forward voltage can be applied, we must ensure that the SCR has turned OFF completely. If the SCR had not turned OFF completely when voltage was reapplied, we would have a switching power loss that would dissipate entirely within the semiconductor and cause excessive heating. In other words, we have a critical dV/dt that severely limits high-speed operation. We can achieve a clearer understanding by referral to Fig. 5–7, where we have the typical dissipation of an SCR during its turn-ON time. A high temperature within the SCR can lead to false triggering and can even prevent commutation.

In SCR-controlled energy converters a simple and very effective solution to commutation is to use resonant inverters, where ringback commutates the SCR for at least a time sufficiently long enough to turn it OFF. The question we need to raise to see some advantage in our power FETs is: How fast, or at what speed, do these SCR resonant

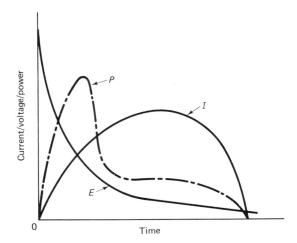

Figure 5-7   Typical power dissipation of an SCR during turn-ON.

inverters operate?   The answer depends, of course, on the size of our inverter.   Since we are obviously interested in using power FETs, our interest would be in a moderate-size inverter, say one limited to 1 kW. Here our answer would probably be to say that the SCR would be operating between 20 and 40 kHz, probably closer to 20 kHz.

We should certainly stop to consider cost.   There are situations where SCRs will always undersell the power FET.   One that comes to mind is the household light dimmer switch, which sells for only a few dollars and is simplicity itself.   Other similar appliances using SCRs or thyristors (such as the Triac) might also be priced too low to become targets for power FETs.

As we argue for the replacement of the power bipolar transistor we can use the same argument to replace the SCR in the moderate power system converter.   A higher switching speed needs less costly and fewer components.   Hopefully, those components that we can eliminate are the expensive ones.   Some indeed are.

5.4 THE BASIC INVERTER

The basic inverter is really little more than a dc to ac energy converter. Used on commercial power, we would have an input rectifier followed by the switching circuit and output transformer.   A block diagram of this elementary circuit is shown in Fig. 5-8.   We can immediately elaborate on the switching element and transformer circuitry showing two inverters in Fig. 5-9.   Here we have both a power bipolar transistor driver and a SCR driver, and we should note the basic similarities between these two circuits.   More often than not, we find that circuits

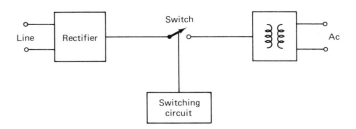

Figure 5-8   Basic inverter building block.

using bipolar transistors often contain reactive elements similar to what we would expect to see in an SCR circuit for commutation. It is important that we interrupt the stored energy found in the highly reactive transformers that might cause destructive second breakdown. A popular energy suppressor is the use of snubber networks, which, as we can see in Fig. 5-9, consist of a series resistor–capacitor network across the primary of the transformer. Also common to both the biploar transistor and SCR is a diode placed in shunt with the switching element. This is very important to "catch" the backwave resulting from the collapsing magnetic field each time the switch opens. Without this catch diode the voltage would literally soar and breakdown would be the unfortunate result.

Generally, our driving circuits for either the bipolar transistor or the SCR are very similar. For either we must guard against the deleterious effects of minority-carrier storage time for the bipolar transistor and commutation for the SCR. Were we to gate either switch with a square wave, we could have serious problems. For the bipolar transistor, storage time could result in both transistors being ON simultaneously, and for the SCRs, insufficient commutation would result in excessive dissipation within the semiconductor that could be destructive. It is not necessary to repeat the problems of high frequency (or high-speed) switching.

By now we should be able to recognize at least two advantages for power FETs. Every power MOSFET has a catch diode that is intrinsic to itself, and, furthermore, we are now well aware that no FET, MOSFET or otherwise, exhibits minority-carrier storage time. Finally, the power FET is a voltage-driven transistor, not current-driven, which greatly simplifies the driving circuitry.

5.5 THE MOSFET-DRIVEN SWITCHER

As we have seen in this and preceding chapters, the power FET offers many outstanding advantages over other power semiconductors, in particular its ability to withstand high-stress conditions and its very fast

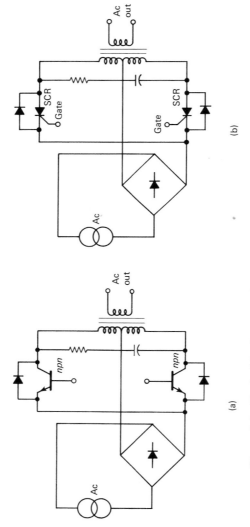

Figure 5-9 Similarities between the (a) bipolar transistor and (b) SCR switching inverters. Note that both require snubbing networks across the transformers and diodes across the switching element.

(a)

(b)

switching speed. Yet there are other more subtle advantages that we need to reemphasize at this time. Power MOSFETs have been exclusively enhancement mode, which simply means that a positive voltage (for *n*-channel) must be applied to turn them ON. Together with this feature is their threshold voltage, which for the high-voltage MOSFETs should be at least 3 V or more. A high threshold voltage will help materially in preventing false triggering, especially in noisy environments such as might be found in inverters and switchers. The power MOSFET, like any FET when used as a switch, is always operated in its triode region (Fig. 2-1), as this provides us with the lowest ON resistance, and thus the lowest voltage drop. As we saw in Chap. 2, the voltage drop is proportional to the total current being passed through the FET.

Because of the high voltages attainable with power MOSFETs, we are able to exercise considerable lattitude in our design philosophy. One popular design approach that offers a low parts count and as a result is often economical for low-power applications is the flyback switching power supply. Although it has the advantage of a reasonably low parts count, it does have the disadvantage of requiring rather heavy filtering and possibly emitting somewhat heavy *electromagnetic interference* (EMI) unless special precautions are taken. The schematic shown in Fig. 5-10 finds excellent utility as a microprocessor power supply by offering 5 V at 10 A. Construction is greatly simplified by the use of several commercially available integrated circuits that handle all the critical areas. Operationally, the output voltage is monitored and compared against a reference using an op amp (RCA3140) which, in turn, controls the light output of the optocoupler used to set the current level, which, in turn, establishes the operating frequency of the 74C14 hex Schmitt inverters driving the gate of the power MOSFET. To assure that the heavy input capacity of the power MOSFET (800 pF) will both charge and discharge quickly enough, an emitter-coupled drive similar to that shown in Fig. 5-2 is used. The flyback transformer uses a ferrite core (for example, a good core material might be Ferroxcube 3C8) for optimum efficiency at high frequencies and we would need to guard that the maximum flux density not exceed the B_{sat} for that core at our highest operating temperature. To ensure that our primary leakage inductance is kept low, we would, in all probability, try to wind the secondary, or output winding, interleaved with the primary. If we are successful, our total power loss might be twice the core loss (copper losses equaling core losses), which we can deduce from the ferrite catalog knowing the mean operating frequency.

We might compare the schematic diagram of our power MOSFET flyback switcher with a commercially available power bipolar transistor

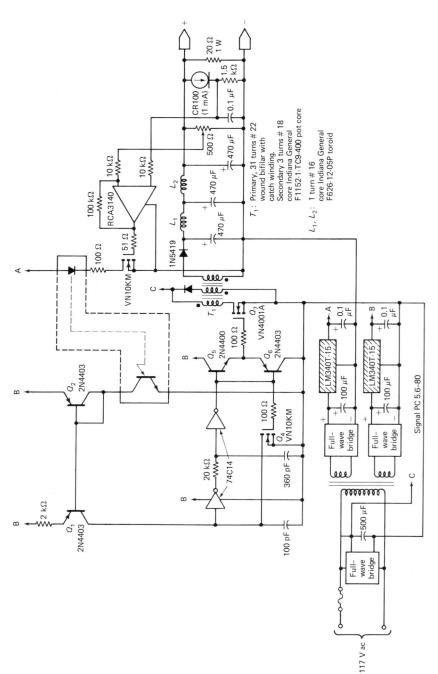

Figure 5-10 A 5-V 10-A flyback regulator using power FETs. (Courtesy of Siliconix incorporated.)

flyback switcher (Boschert, Inc., model OL25) and note the obvious similarities. This we can do by comparing Fig. 5-10 with Fig. 5-11. In the bipolar transistor flyback switcher, feedback is accomplished using an optocoupler to control the blocking oscillator, Q_1. Both duty cycle and power vary according to need, somewhat akin to the operation of the power MOSFET flyback switcher.

Using power MOSFETs in a totem-pole arrangement we can easily assemble a half-bridge or a full bridge switcher. One of the most efficient forms for driving power MOSFETs (as well as power bipolar transistors) is to use *pulse-width modulation* (PWM). Rather than offering variable-frequency drive, as we did for the flyback switcher, which for a low parts count was desirable, we have fixed-frequency drive and pulse-width modulation. That is, our ON and OFF times are variable. Pulse-width modulation provides a sinusoidal wave which is principally dependent upon the number of pulses and their spacing, as shown in Fig. 5-12. A PWM switcher also offers us greater flexibility in power handling. Whereas the flyback switcher was excellent at low power (up to 50 W), the half- and full-bridge circuits offer power limited only by the current-handling capability of the switching transistors. Here is another bonus for using power MOSFETs. Were we to increase the current handling of power bipolar transistors, their storage time would stretch out and would severely limit the switching time, even with antisaturation circuits, as shown in Fig. 5-1. It is important that when paralleled power transistors switch OFF, they switch in unison; otherwise, we would have excessive switching losses due to having a voltage across the transistor as well as current flow through the transitor simultaneously, with a result similar to that shown in Fig. 4-30. As we have seen, the power MOSFETs have unique paralleling capability, partially due to their degenerate thermal characteristics if we only try to match threshold voltages to within, say, 10%, and take the usual precautions against parasitic oscillations (Fig. 5-6). We should take special note of another advantage of the totem-pole bridge design. To ensure optimum *brownout* operation, that is, to ensure that we can, indeed, deliver rated voltage and current when the line is interrupted by a momentary disturbance, we should attempt to design our switchers to operate "off line" without the need for voltage-reducing transformers. Most switchers that have been designed to operate "off line" can do so with a typical line voltage fluctuation of 90 to 130 V ac. Our brownout insurance stems from the input filter capacitor following our initial ac to dc conversion. The charge held by this filter capacitor is generally sufficient to maintain the status quo for at least a couple of missed cycles. A low-voltage filter capacitor would have to be unduly large, which would

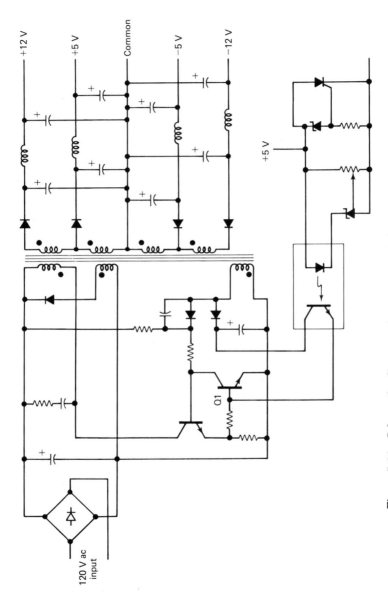

Figure 5-11 Schematic diagram of a commercial flyback switching converter using power bipolar transistors. (Reprinted with permission of *Electronics*, Dec. 21, 1978, © McGraw-Hill, Inc., 1978. All rights reserved.)

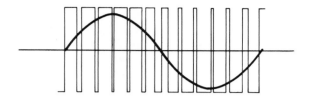

Figure 5-12 Sinusoidal waveform generated from a pulse-width modulator (PWM).

mitigate against its use in an efficient and economical switcher. The totem-pole bridge design also offers practical use of switchers in foreign service, where line voltages run at 220 V ac. Of course, to use power MOSFETs in a totem-pole arrangement for either line voltage, we need high-breakdown FETs.

We should be aware that to achieve efficient switching using the totem-pole concept, we must ensure that our power MOSFETs have exceptionally low ON resistance, certainly equal to or less than 1 Ω. Otherwise, we might be fooling ourselves into thinking that our switcher is efficient! We have the advantage, of course, that as our power needs increase, we might parallel the MOSFETs, which will drive the ON resistance down. The immediate concern involves saturation voltages, which for the power bipolar transistor are generally low. For the power MOSFET, saturation voltage can be calculated from a knowledge of its ON resistance, $R_{DS(on)}$ *times* the drain current, I_D. If we allow ourselves to be caught up entirely in such comparisons, we might fail to see that the overall benefits we gain from our use of power FETs far outshines the singular problem of $V_{CE(sat)}$, whether it be of a bipolar transistor or MOSFET. Part of our argument in favor of the power MOSFET stems from the facts presented in Table 4–1, where we found that when drive requirements and switching losses are considered, power MOSFETs generally outperformed comparable power bipolar transistors. If we were to accept the concept of an equal-power-frequency figure of merit, we would find that as the switching frequency rises, power bipolar transistor efficiency decays and power MOSFET efficiency rises. Our basic premise for efficiency was not merely power in versus power out, but also included MTBF and costs, and switchers using power MOSFETs definitely offer improvement if operated at a high switching frequency.

Both the half-bridge and the full-bridge switchers are to be generally preferred over most push-pull switchers, for at least two reasons. The most obvious that we can immediately recognize from our previous familiarity with the totem-pole arrangement is that each FET does not need to withstand the full operating voltage. The second reason is not as obvious, but may, indeed, be even more important. As we lower our breakdown-voltage requirements for the FETs, we would expect, from our understanding of Eq. (4.13), to witness a dramatic reduction in the MOSFET's ON resistance. In fact, we would see such a reduction in ON resistance that we would experience a marked improvement in over-all efficiency. An added benefit of low ON resistance is lower power loss ($I^2 R_{DS(on)}$) and, as a direct result of this, a lower temperature rise of the power MOSFET. Since the bulk resistivity of silicon exhibits a positive temperature coefficient of resistance—in the approximate range 0.6 to 0.8% per °C—a low-temperature rise is very definitely to our benefit. Furthermore, we would be wise to remember the *negative* co-efficient for threshold voltage with increasing temperature of approx-imately—5 mV per °C. A high operating temperature from any cause might so reduce our threshold voltage as to make our power MOSFETs sensitive to noise transients, which could lead to false triggering and potential catastrophic destruction of both the power MOSFETs and possibly the entire switcher.

We need not be reminded of the importance of an adequate heat-sink; if we are familiar with using power bipolar transistors, we have un-doubtedly had experience. However, because of a power FET's degen-erate thermal characteristic it is important that we review some basic fundamentals so that we can be assured of long life and, with it, opti-mum reliability. Since the life and reliability of any power semiconduc-tor is directly related to its maximum junction temperature, we can appreciate that our task is twofold. First, we have got to determine the junction temperature, and, second, to design or establish the size of our heat sink that will ensure that the junction temperature will not be exceeded under normal operation.

Our first task is to determine the total power dissipated in the power FET. We might at first believe that if we "lifted" $R_{DS(on)}$ from the data sheet and, knowing the drain current, apply Ohm's law, we would arrive at the correct value. This would certainly be handy *if* our FET did not have a temperature-dependent ON resistance. We learned from earlier chapters that all bulk semiconductors, such as our power FETs, have ON resistances that increase with increasing temperature, and regrettably, this increase is *not* linear, nor is it the same for all power FETs!

In any power FET application where the current is defined by the load, such as in switching power supplies or motor controllers, we must be especially careful to guard against *thermal runaway*. We might at first argue that a thermally degenerate power FET is safe from thermal runaway, but let us take a closer look. We saw in earlier chapters that $R_{DS(on)}$ increases with increasing temperature, and as a consequence it is easy for us to parallel FETs. In fact, that is one of the benefits of power FETs that makes their manufacture easier than bipolars; we can forget ballast resistors. However, if the drain current is controlled (or defined) by some external force (or load) and our FET heats up, *the I^2R losses increase*! In other words, we have what amounts to thermal runaway that can be catastrophic. It may have all the appearances of secondary breakdown and we may erroneously believe that is what happened, but although the results are the same (catastrophic destruction), the event was, in reality, thermal runaway.

Since we are considering the use of power FETs in switchers, let us consider the steps necessary for an adequate heatsink. Irrespective of application, if the drain current is load-defined, we will need a clear understanding of the following discussion.[1]

The total power that we need to dissipate in a heat sink can be determined from the equation

$$P_T = P_S + P_C + P_G + P_L \tag{5.1}$$

where P_T = total power to be dissipated

P_S = switching transition loss

P_C = conduction loss

P_G = drive power lost in the gate

P_L = leakage (I_{DSS}) during the OFF period

Since power FETs have no minority-carrier storage time, we should be able, by careful design, to reduce switching transition losses P_S to negligible amounts. However, such is often not the case. In Fig. 4-30 we saw the switching waveform for an inductive load, and if we were to allow the enclosed area of dissipative power to resemble a triangle with straight sides [that is, I_D(fall) and V_{DS}(rise)], we could then calculate P_S as

$$P_S = \frac{V_{DS(\text{max})} \times I_{D(\text{max})}}{2} \ \tau_s f_s \tag{5.2}$$

where f_s is the switching frequency.

One interesting anomaly that we must be prepared to consider is the stray output capacity in parallel with C_{oss} resulting from insulating the FET from the heatsink. A mica or thin-plastic-film insulating washer for a TO-3 will add nearly 200 pF and a 0.062-in. (1.57-mm)

---

[1] The following discussion on power dissipation and junction temperature has been condensed from an application note authored by Rudy Severns of International Rectifier, Inc., entitled, "Simplified HEXFET Power Dissipation and Junction Temperature Calculation Speeds Heatsink Design," *International Rectifier Application Note 942.* Used with permission.

beryllium oxide washer, 25 pF. Figure 5-13 graphically shows us the debilitating effect upon turn-OFF time that results from a 200-pF plastic-film insulating washer. If we again referred to Fig. 4-30, we would see an altered V_{DS} rise time.

The power lost through leakage (I_{DSS}) is generally insignificant, even when we have our junction temperature at the maximum allowable level.

Gate drive losses P_G depend to a great extent upon whether or not our MOSFET has a polysilicon or metal gate. Newer power MOSFETs (not used for high-frequency RF applications) are generally designed with polysilicon gates, and at high switching speeds we might anticipate some gate losses. Generally, however, these losses are insignificant compared to the conduction losses.

Without any doubt, as we saw from our examination of Fig. 4-30, the conduction loss P_C is the major power loss for a power FET. We can identify this loss using the simple equation

$$P_C = I_D^2(\text{rms}) \times R_{DS(\text{on})} \tag{5.3}$$

If we wish to use this equation effectively, we have got to be able to determine the rms value of I_D irrespective of waveform. The rms value of any waveform is defined as

$$I_{\text{rms}} = \left\{ \frac{1}{T} \int_0^T [I(t)]^2 \, dt \right\}^{1/2} \tag{5.4}$$

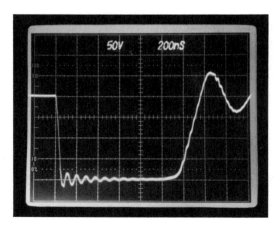

Figure 5-13 Typical switching waveform of a power MOSFET packaged in a TO-3 and mounted on a heat sink using a 0.003-in. mica washer that contributes nearly 200 pF in shunt with the MOSFET's C_{oss}. Note that only t_{off} is affected. (Courtesy of International Rectifier, Inc., Semiconductor Division.)

Equation (5.4) has been simplified for our practical application in Table 5-1.

We can now begin to zero-in on our thermal-dissipation and heat-sink problem by establishing a graphical solution using available graphs normally found in any power MOSFET data sheet. If we conclude that P_C is the major power loss, we can group the remainder as follows:

$$P_1 = P_S + P_G + P_L \tag{5.5}$$

Combining Eqs. (5.5) and (5.3) into Eq. (5.1), we get

$$P_T = P_1 + I_{D(\text{rms})}^2 R_{DS(\text{on})} \tag{5.6}$$

We can also derive a simple equation that will provide us with the junction temperature

$$T_J = T_A + R_{\Theta_{JA}} P_T \tag{5.7}$$

where T_A is our ambient temperature, $R_{\Theta_{JA}}$ the thermal resistance between the semiconductor junction (active area) and the ambient, and P_T is the dissipated power from Eq. (5.6). $R_{\Theta_{JA}}$ actually consists of several *serial* thermal resistances: $R_{\Theta_{JC}}$ being the thermal resistance between the chip and the package (such as a TO-3), $R_{\Theta_{CH}}$ the resistance between the case and the heat sink, and $R_{\Theta_{HA}}$ the resistance between the heat sink and the surrounding ambient. The equation being

$$R_{\Theta_{JA}} = R_{\Theta_{JC}} + R_{\Theta_{CH}} + R_{\Theta_{HA}} \tag{5.8}$$

If we are faced with a design problem requiring a heat sink (and there are precious few situations where we won't be!), we should select our power FET from a vendor whose data sheet provides a normalized $R_{DS(\text{on})}$ versus junction-temperature (T_J) graph. Since the successful solution of our problem requires the simultaneous solution of both P_T and T_J [Eqs. (5.6) and (5.7)], a graphical approach is *much* easier.

To prepare our equations for this graphical solution we begin by normalizing both P_T and T_J. Since Eq. (5.6) holds for *any* temperature, we can rewrite it as

$$P_{T(25°\,C)} = P_1 + I_{D(\text{rms})}^2 R_{DS(\text{on})25°\,C} \tag{5.9}$$

Combining Eqs. (5.6) and (5.9), we obtain the normalized equation

$$P_N = \frac{P_T - P_1}{P_{T(25°\,C)} - P_1} = R_{DS(\text{on})N} \tag{5.10}$$

What this allows us to do is to relabel our normalized $R_{DS(\text{on})}$ versus T_J curve (found in the data sheet) as P_N versus T_J. But before we

TABLE 5-1 Effect of waveform shape on RMS value of current. [Used with permission from International Rectifier Application Note AN-942.]

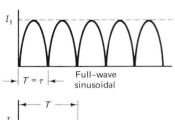

Full-wave
sinusoidal

$$I_{RMS} = \frac{I_1}{\sqrt{2}}$$

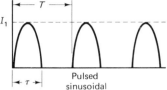

Pulsed
sinusoidal

$$I_{RMS} = I_1 \sqrt{\frac{D}{2}}$$

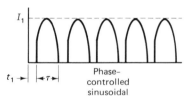

Phase-
controlled
sinusoidal

$$I_{RMS} = I_1 \left[\frac{D}{2} + \frac{\sin \tau (1 - D) \cos \pi (1 - D)}{2\pi} \right]^{1/2}$$

$$D = 1 - \frac{t_1}{T}$$

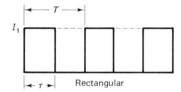

Rectangular

$$I_{RMS} = I_1 \sqrt{D}$$

$$D = \frac{\tau}{T}$$

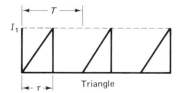

Trapezoid

$$I_{RMS} = D \left(\frac{I_a^2 + I_a I_b + I_b^2}{3} \right)^{1/2}$$

$$D = \frac{\tau}{T}$$

Triangle

$$I_{RMS} = I_1 \sqrt{\frac{D}{3}}$$

$$D = \frac{\tau}{T}$$

continue we must normalize T_J of Eq. (5.7):

$$T_J = T_A + R_{\Theta_{JA}} [P_1 + P_N (I_D^2 (\text{rms}) R_{DS(\text{on})25°C})] \qquad (5.11)$$

and

$$P_T = P_1 + P_N (I_{D(\text{rms})}^2 R_{DS(\text{on})25°C}) \qquad (5.12)$$

Once we determine P_N we are able to read T_J directly and, using Eq. (5.12), calculate P_T. Having followed these derivations, the implementation is straightforward, using the normalized $R_{DS(\text{on})}$ versus T_J curves supplied on the power FET data sheet.

Before we use these normalized equations in an example, there is a small matter that needs our attention which we touched only briefly in Eq. (5.8). Before we can use $R_{\Theta_{JA}}$ in our equation we must know $R_{\Theta_{CH}}$ and $R_{\Theta_{HA}}$. $R_{\Theta_{JC}}$ is given on the data sheet. The thermal resistance between case to heat sink, $R_{\Theta_{CH}}$, is available from numerous sources, some of which are contained in the References at the end of the chapter. Power FETs, for the most part, have been packaged in either the TO-3 or the TO-220 and, for convenience the thermal resistance for these two packages is provided in Table 5-2.

TABLE 5-2 Interface Thermal Resistance $R_{\Theta_{CH}}$ Using Thermal Grease		
Insulator	TO-3	TO-220
0.002-in. Thermafilm	0.52	2.25
0.003-in. mica	0.36	1.75
0.02-in. anodize	0.28	1.25
0.062-in. BeO$_2$	0.18	—
No insulator	0.14	1.0

The thermal resistance between the heat sink and ambient, $R_{\Theta_{HA}}$, is provided by the manufacturer of the heat sinks. Be sure that you have the value before you make your purchase!

Let us follow an example using the International Rectifier HEXFET IRF431 and graph our solution on the data sheet-supplied normalized resistance curve, as shown in Fig. 5-14.

The IRF431 HEXFET data sheet provides us with some basic information.

Case	TO-3
$R_{\Theta_{JC}}$	1.7°C/W
$R_{DS(\text{on})25°C}$	1.7 Ω at $I_D = 2.1$ A

Let us assume our ambient T_A at 55°C, and have our rms drain current $I_{D(\text{rms})} = 2.1$ A. For our illustration we can assume the various minor losses to be insignificant, so $P_1 = 0$.

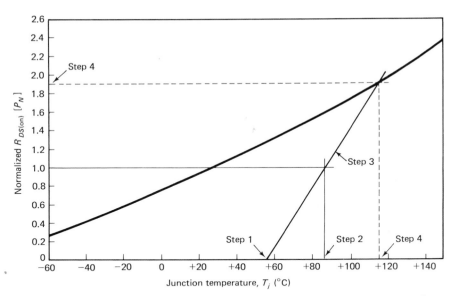

Figure 5-14 Normalized $R_{DS(ON)}$ versus junction temperature of IRF431. (Courtesy International Rectifier, Inc., Semiconductor Division.)

Using a mica washer (see Table 5-2) with thermal grease, we have $R_{\Theta_{CH}} = 0.36°C/W$. Our heat sink has an advertised thermal resistance $R_{\Theta_{HA}}$ of 2°C/W.

Step 1: If we have zero power through the HEXFET, our junction temperature T_J equals the ambient T_A. Locate our first point for

$$P_N = 0 \qquad T_J = T_A = 55°C$$

Step 2: Next locate our second point for $P_N = 1$ and use Eq. (5.11) to calculate T_J:

$$T_J = 55°C + (1.7 + 0.36 + 2)(2.1^2)(1.7) = 85.4°C$$

Step 3: Draw a straight line through

$$P_N = 0, T_J = 55°C \qquad \text{and} \qquad P_N = 1, T_J = 85.4°C$$

and extend the line to *intersect* the normalized $R_{DS(on)}$ curve.

Step 4: Read T_J and P_n directly.

$$T_J = 115°C \qquad P_N = 1.9$$

Step 5: Calculate P_T from Eq. (5.12):

$$P_T = 0 + (2.1^2)(1.7)(1.9) = 14.24 \text{ W}$$

Returning to our totem-pole bridge switchers, we have now resolved how to maintain the junction temperatures of our power MOSFETs within their safe operating specification and we have also developed an appreciation for why we need low ON resistance FETs. Heat loss is loss of efficiency. A power supply that does not require heat sinks is certainly more efficient than one that does. Heat loss is only one reason why switchers are more efficient than linear power supplies.

An elegantly simple half-bridge switcher is the Hoffman switcher shown in Fig. 5-15, where a pair of 400-V power MOSFETs arranged in a totempole are triggered from a commercial integrated circuit designed specifically for PWM switching power supplies that allows operation at 100 kHz (the design will allow operation at much higher frequencies). As with the flyback design, the emitter-coupled bipolar transistor pairs provide sufficient gate current to more than adequately charge and discharge the gate–source capacity in the time required. Driving power MOSFETs arranged in a totempole is easy, as we see here. The 50 Ω series gate resistors help prevent unwanted parasitic oscillation. T_3 and D_{15} provide automatic shutdown should the drain current through the totem-pole pair of power MOSFETs rise to what might be dangerous levels, which, incidentally, can be set by R_{24}. A very worthwhile design feature highly recommended not just for switchers but for many designs is the soft startup thermistor, TH_1.

Since feedback is taken from the +5-V output, we must properly assume that only the +5-V supply is fully regulated, all others being quasi-regulated. A simple expedient that provides good regulation at moderate to low output power is to use a pair of voltage regulators, as shown and identified as IC_4 and IC_5.

There are at least two critical areas in the design of any high-speed switcher: the circuit board and the main power transformer fed by the switching transistors, which in our case are power MOSFETs. Because of the high switching speed it is important that we construct the circuit board as we might for a high-frequency design using *double*-copper-clad Teflon laminate, or some other high-quality material. Grounds are especially critical and plated-through holes preferred. Short-lead, point-to-point wiring is mandatory and reasonable separation should be allowed where pulse leads may run an appreciable distance. Transformer design is even more critical and the interested reader should consult the References at the end of the chapter. Of special importance is the problem of EMI, which in recent years is coming under increasing scrutiny and is resulting in legislation limiting its effect. Switchers are notoriously noisy and as new designs are possible with power FETs, their switching frequency is on the rise and could (or does) cause serious interference to both military and maritime communications as well as broadcast.

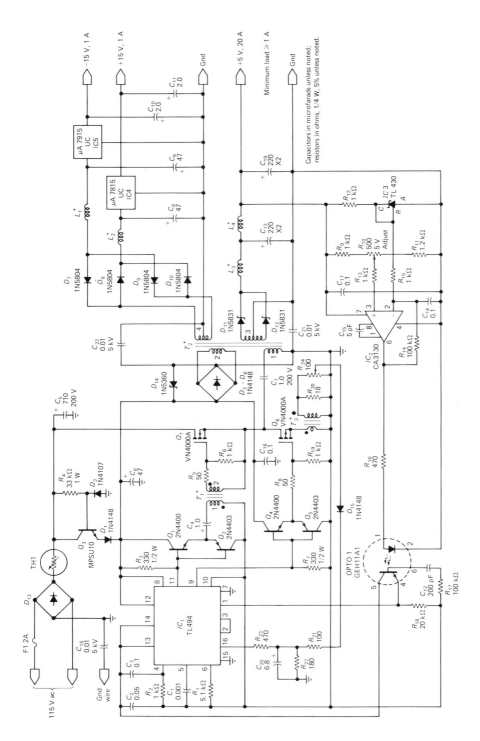

Figure 5-15 A 100-kHz 150-W half-bridge switching power supply using 400-V power MOSFETs. (Courtesy of Siliconix incorporated.)

A full-bridge switcher can evolve directly from our totem-pole half-bridge circuit by using a twin totem-pole arrangement, as shown in Fig. 5-5.

5.6 THE BASIC REGULATOR

The regulator differs from the switcher since it acts as a regulator of poorly filtered (or more probable poorly regulated) dc to produce highly regulated dc for critical applications. Often they are recognized as series regulators, that is, where the power transistor is a series element. The basic regulator is shown in Fig. 5-16. Regulation is accomplished by sensing the output voltage and comparing it to a reference. The error signal is then used to control the series pass regulator transistor, either directly or by PWM.

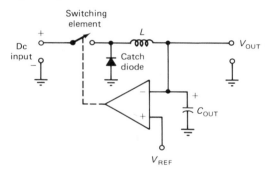

Figure 5-16   Basic switching regulator.

As we refer to Fig. 5-16, we can follow its operation by first beginning our examination with the series switch closed. Direct current will flow from our source through the series inductor, charging the filter capacitor and out to the load. If we were to open the switch, we would, quite obviously, disconnect our source of current from the load. Immediately, the electromagnetic field about the inductor will collapse, but in so doing will generate a current. This current path will loop through the load and back via the catch diode in shunt across the regulator. If we now place a voltage-sensing circuit across the output that turns the switch ON once a voltage droop is sensed, we have, in essence, a voltage regulator. The degree of regulation, or ripple, due to the periodic voltage droop is dependent upon both the sensitivity and slew rate of our feedback network. The shunt filter capacitor helps smooth the ripples caused by the switching action.

In Fig. 5-17, a Siliconix VMOSFET, type VN64GA, is used as the series switching element in a regulator particularly suited for microprocessor applications, providing a well-regulated 5-V 10-A output with

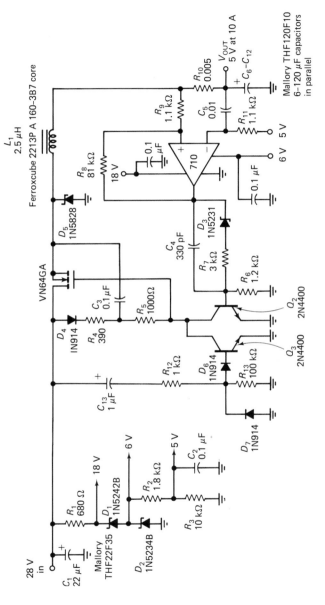

Figure 5-17 A 200-kHz switching regulator using the Siliconix VN64GA as a series switching element. (Courtesy of Siliconix incorporated.)

less than 100 mV of ripple. The basic elements of this circuit follow those of Fig. 5-16 with few exceptions, which we examine presently.

If we begin with the source of power, we have a raw 28-V supply from which we take from a bleeder the necessary operating voltages, 5 V, 6 V, and 18 V. A very beneficial safety feature has been designed into the regulator, commonly called a *soft startup*, consisting of C_{13}, R_{12}, and Q_3, which clamps the series VMOS switch hard OFF until C_{13}, reaches a full charge (suggesting that the supply has stabilized). The LM710 comparator is actually an oscillator whose frequency is controlled by L_1 and R_8, which in our case as shown here runs at 200 kHz. The output of the comparator shifts down by means of the level shifter, consisting of zener, D_3, and resistors R_6 and R_7, to provide control to the base of the *npn* bipolar transistor, Q_2. In order that we may turn the VMOSFET full ON we know that the gate-to-source voltage must be some positive value with respect to the source. Since it is a series pass transistor under normal operation, the source will be at the highest potential anywhere on the board. For us to raise the gate voltage to be more positive than the source, we must use what is known as a *bootstrap* circuit.

The bootstrap circuit is an important adjunct to the successful operation of enhancement-mode power MOSFETs. We find them used in circuits where we need to *switch* stacked, or totem-pole, arrays of power MOSFETs. It is important that we immediately recognize that the bootstrap circuit is useful *only* for switched, pulsed, or ac circuits. The reason will soon be obvious. A basic bootstrap circuit is shown in Fig. 5-18. In this basic circuit we wish to provide a gate-to-source voltage sufficient to turn the power MOSFET full ON. That means we need a substantial positive gate-*to-source* voltage. When Q_1 is OFF, Q_2 is ON and the gate of Q_1 is at near ground potential. C_1 charges to full voltage less the drop across D_1 and R_L. To turn Q_1 ON we turn Q_2 OFF, which allows the gate of Q_1 to rise. However, the gate does not

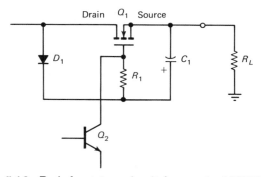

Figure 5-18   Basic bootstrap circuit for a series MOSFET switch.

merely rise to the supply voltage (less the diode, D_1, drop) but *takes the added voltage stored in C_1*, back-biasing D_1, and turning Q_1 hard ON. By the time C_1 discharges through the parallel back-biased diode and the high-resistance gate, the next switching pulse begins the procedure over again.

The benefits resulting from our 200-kHz voltage regulator are by now obvious. Aside from smaller and perhaps fewer components, we find that the response time improved by an order of magnitude over what we would expect from a bipolar transistor regulator operating at one-tenth the frequency.

With the large selection of available power FETs we can modify the basic regulator circuit to suit numerous applications. We should be careful to use an adequate heat sink for the series pass MOSFET.

5.7 THE CURRENT REGULATOR

In Chap. 2 we found that all short-channel power MOSFETs, as well as all small-signal planar J-FETs, exhibit current saturation beyond a prescribed drain–source voltage limit. For the small-signal J-FET this occurred at voltages beyond pinch-off; for the short-channel power MOSFET it occurred at velocity saturation. In Fig. 2–1 we called this region the pentode region. If we could bias the gate of a power MOSFET to ensure that the drain current would saturate, we would have, in effect, a constant-current source. The principal advantage of using a power FET as a constant-current regulator is its very high output impedance ($1/g_{os}$), which provides exceptionally stable current over wide voltage fluctuations. Additionally, unlike the bipolar transistor, we see little sensitivity to temperature variations. Unlike the switchers we have just studied, where the power MOSFETs operated in their triode, or low-$R_{DS(on)}$ region, here we will use these power MOSFETs in their saturation, or high-$R_{DS(on)}$ region and so must be cautious of excessive temperature rise during operation. An adequate heat sink will be very important to the success of our particular application involving current regulators. As we progress through this section we will find power MOSFETs ideal as either current sources or current sinks. The latter is especially useful as a constant-current load and with some ingenuity can be designed for digital current control using little more than a digital-to-analog (D/A) converter to supply the reference voltage.

The basic circuit, shown in Fig. 5–19, is for the most part the actual operating circuit for a current sink and the rudiment for a current regulator. The operation of this circuit is quite simple. A power MOSFET capable of withstanding the maximum voltage and handling the maximum anticipated current is first selected and a low-ohmic, high-wattage resistor is used as the sense element. An op amp makes a

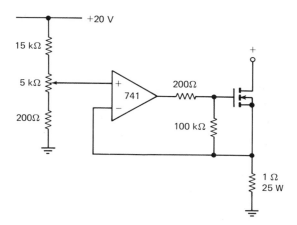

Figure 5-19 High compliance current load. (Courtesy of Siliconix incorporated.)

convenient comparator monitoring the voltage across the sense resistor and comparing it with our reference, which can be a potentiometer, a zener, or a D/A converter. The accuracy of our circuit depends almost entirely upon the stability of the sense resistor and the offset of the op amp, and, of course, on the stability of our reference voltage.

The circuit shown in Fig. 5–20 offers either a current sink (a) or a

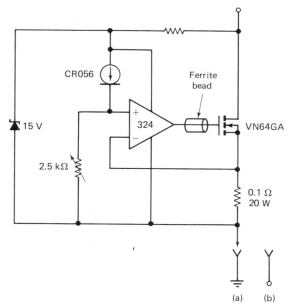

Figure 5-20 Current sink (a) or two-terminal regulator (b). (Courtesy of Siliconix incorporated.)

two-terminal regulator (b). Using the Siliconix VN64GA allows us the latitude of current regulation [in Fig. 5–20(b)] from a few milliamperes to the maximum allowed by the FET.

5.8 SUMMATION OF BENEFITS AND PROBLEMS
IN USING POWER MOSFETs

As we have progressed through this chapter, we focused much of our attention on the rather obvious benefit that all FETs offer; no minority-carrier storage time, which made it possible for us to increase the switching speed. The side benefits were several. First, we needed fewer and less expensive parts; this, in turn, helped improve reliability. A higher-speed switcher provided faster response time to fluctuating load conditions. Because of the degenerate thermal characteristics of the power FET, we found that with a much improved safe operating area it was more difficult to stress the transistor, and as a consequence our power FET was more reliable. In Chap. 4 we determined that power FETs are generally more efficient switchers, in that no drive power is needed in comparison to most bipolar transistors.

But there are disadvantages. Perhaps the two major disadvantages that have captured the attention of critics and users alike are, first, the high $R_{DS(on)}$ resistance, which effectively increases the $V_{DS(on)}$ of a power FET over the V_{SAT} of a comparably rated power bipolar transistor. A related problem is the temperature coefficient of the bulk silicon, which causes the ON resistance to rise more quickly than does the V_{SAT} rise for a power bipolar transistor. For example, $R_{DS(on)}$ will double when the chip temperature reaches $150°C$.

The second major disadvantage of some concern is the maximum permissible operating temperature generally found on all power MOSFET data sheets: $150°C$. Silicon power bipolar transistors are nearly always rated to $200°C$. We can appreciate the problem if we consider the maximum power available for a transistor derated from $150°C$ as compared with a similarly rated power bipolar transistor derated from $200°C$. That $50°C$ difference makes the derating problem critical in many applications, such as under-the-hood automotive and many military systems, where ambient temperatures often are close to $100°C$. Coupled with this thermal problem is the root cause, which is a decreasing threshold voltage with increasing chip temperature. What compounds our problem is that the biasing conditions change with temperature. What might be an OFF condition at $25°C$ we may find ON at $100°C$, or at least we may find ourselves needing to review our biasing circuitry.

Generally, we regard the degenerate thermal characteristics a very worthwhile asset, but there is a situation where great caution must be exercised. If we use a power FET in a constant-current mode, such as in one of the current sinks that we discussed in this chapter, we must guard against temperature rise, which, in turn, raises the bulk resistivity. As the resistance increases we should immediately recognize what is about to happen. A constant current I_D and increasing channel resistance $R_{DS(on)}$ results in rapidly increasing power dissipation $I_D{}^2 R_{DS(on)}$, which means that we see a dramatic increase in temperature. The final effect, if left unchecked, appears to the uninitiated as second breakdown! Heat sinks are important and we cannot emphasize their use too strongly.

Nonetheless, despite these disadvantages, power FETs have found an ever-widening role in applications requiring high-speed switching and, perhaps surprising because of their apparent disadvantages, high reliability.

REFERENCES

BOSCHERT, ROBERT J., "Flyback Converters: Solid-State Solutions to Low-Cost Switching Power Supplies," *Electronics*, Dec. 21, 1978, pp. 100–104.

CLEMENTE, S., B. PELLY and R. RUTTONSHA, "A Universal 100 kHz Power Supply Using a Single HEXFET™," *International Rectifier Application Note 939*, El Segundo, Calif., International Rectifier Semiconductor Division, 1981.

DOYLE, JOHN, M., *Digital, Switching, and Timing Circuits*, North Scituate, Mass.: Duxbury Press, 1975.

HAVER, ROBERT J., "The ABC's of DC and AC Inverters," *Motorola Application Note AN-222*. Phoenix, Ariz.: Motorola Semiconductor Products Division, 1972.

HIRSCHBERG, WALTER J., Sessions Chairman, "The Future of Switching Power Supplies," Session 34, ELECTRO, New York, 1979.

JACKSON, HERBERT W., *Introduction to Electric Circuits*, 4th ed. Englewood Cliffs, N.J.: Prentice-Hall, Inc., 1976. An excellent treatise on the subject of transformers useful in switcher design.

Magnetics Staff, "Inductor Design in Switching Regulators," *Power-conversion International*, 6 (Jan.–Feb. 1980), 45–47. An excellent source for transformer design information.

MALONEY, TIMOTHY J., *Industrial Solid-State Electronics*. Englewood Cliffs, N.J.: Prentice-Hall, Inc., 1979.

OXNER, ED, ed. *VMOS Applications Handbook*. Santa Clara, Calif.: Siliconix incorporated, June 1980.

PIVIT, ERICH, and J. SAXARRA, "Upgrade Your Switchers Analytically," *Electronic Design*, 26 (May 10, 1978), 108–13.

ROEHR, BILL, "Mounting Techniques for Power Semiconductors," *Motorola Application Note AN-778*. Phoenix, Ariz.: Motorola Semiconductor Products Inc., 1978.

ROEHR, WILLIAM D., ed., *Switching Transistor Handbook*. Phoenix, Ariz.: Motorola Semiconductor Products Division, 1963.

ROEHR, BILL, and BRYCE SHINER, "Transient Thermal Resistance—General Data and Its Use," *Motorola Application Note AN-569*. Phoenix, Ariz.: Motorola Semiconductor Products Inc., 1972.

SEVERNS, RUDOLF, "The Design of Switchmode Converters above 100 kHz," *Intersil Application Bulletin A034*. Cupertino, Calif.: Intersil, Inc., 1980. An excellent treatise on the subject of high-frequency switching power supplies, rectifiers, and filters.

SLOANE, T. H., H. A. OWEN, JR., and T. G., WILSON, "Switching Transients in High-Frequency High-Power Converters Using Power MOSFETs," *Proceedings* Power Electronics Specialists Conference (PESC), San Diego, Calif., June 1979.

ZOMMER, NATHAN, "Designing the Power-Handling Capabilities of MOS Power Devices," *IEEE Trans. Electron Devices*, 27 (1980), 1290–96. A careful study relating to the die attach of power MOSFETs to improve both thermal fatigue and reliability.

six

Using the Power FET in Motor Control

Motor control is speed control. Ever since the introduction of the thyristor in the mid-1950s, dc motor control has captured the major share of the SCR and Triac market.

Dominant in this field is the dc motor ranging from fractional horsepower (hp) to thousands of hp. We find these motors in portable household tools and utensils as well as in massive industrial equipment. The subway and streetcar are powered by variable-speed dc motors as are our common food blenders and electric drills.

In the early days of motor speed control, before thyristors, we saw carbon-pile rheostats reduce motor speed by reducing the armature voltage. To increase the motor speed beyond its normal speed, the carbon-pile rheostat was used to reduce the field current. Among the several problems we had with this control system, other than its awkwardness and lack of remote electronic control, was that our dc motor heated up rather excessively and often our torque would fall dramatically. Other forms of early motor speed control consisted of fluid control systems and various forms of mechanical couplings. In the 1930s electron drives using gaseous-discharge ignitrons were popular to control large industrial motors.

Thyristors opened novel ways for us to vary speed and maintain torque. Vital to the reliability of the dc motor was the need to limit the armature current at low speed, and the thyristor made this possible. Speed could be controlled by voltage, torque was controlled by current. Both were controlled by the adjustment of the *firing angle* of the thyristor. This firing angle accomplished two things: we were able to adjust the armature voltage, hence control the speed, and we were able to control the armature current, which, in turn, controlled the torque. In Fig. 6-1 we have a typical performance chart showing the effect of firing angle upon both speed and torque of a series-wound dc motor.

We find SCRs very attractive for large dc motor control, especially from ac power, as the SCR acts as its own rectifier. Furthermore, SCRs can handle high surge currents and withstand very high voltages. But there are disadvantages. Although perhaps not a problem on ac mains of 50 or 60 Hz, commutation is necessary. We must turn the SCR OFF between every discrete pulse. Another basic problem involves high-speed switching of heavy currents and elevated voltages. An abrupt OFF condition would see our current quickly drop and the voltage across the SCR quickly soar. We might see an abnormal $\partial V/\partial t$ and $\partial I/\partial t$, which, in turn, could result in false triggering. If we were controlling heavy dc motors of high horsepower, such fast turn-OFF might result in abnormally high internal heating of the thyristor.

In our study we shall find that for small dc motors the power

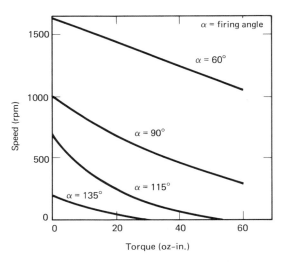

Figure 6-1 Speed versus torque of a fractional-horsepower series motor at various half-wave thyristor control levels. (© 1972 IEEE. Reprinted, with permission from "Solid-State Motor-Speed Controls," by Alexander Kusko, which appeared in *IEEE SPECTRUM*, Vol. 9, No. 10, Oct. 1972.)

MOSFET makes an attractive alternative to the ubiquitous SCR. For heavy industrial motors of hundreds of horsepower, we may eventually see some penetration by power FETs, but certainly not until the power FET can safely and *continuously* handle tens of amperes at high stand-off voltages.

Although we have witnessed considerable progress over the past decades in dc motor control, the dream of designers has been to replace the expensive dc motor with an induction motor. Dc motors are expensive and have reliability problems resulting from their commutators. The squirrel-cage induction motor can be built to be more rugged, and for an equivalent horespower it is smaller than the dc motor. There is basically only one fundamental problem with the induction motor that has restricted its use: its speed is insensitive to voltage. The speed of a synchronous motor is determined by the number of field (or stator) poles and the frequency:

$$\text{rpm} = \frac{120f}{P} \qquad (6.1)$$

As is quite obvious from Eq. (6.1), speed is best controlled by frequency control. Synchronous motors are generally favored when we wish to control speed precisely, or, in some applications, when we wish to interact with other motors in exact synchronism.

In this chapter we shall find that the power FET offers the designer of motor controllers a viable option. We examine and compare thyristors and bipolar transistors (including Darlingtons) with MOS-FETs for a variety of motor control functions. These include dc motors and brushless dc (stepping) motors. We shall also focus our attention on the benefits of using power FETs to control synchronous single and multiphase ac motors. We should be quick to appreciate the fundamental difficulty of using SCRs to achieve speed control in ac motors. Most ac motors are designed for a sine-wave input, and without needless digressing into harmonic analysis we should be able intuitively to visualize that such a sine wave would be, at best, difficult or even impossible to achieve using SCRs.

6.2 DC MOTOR CONTROL

If we had been denied the benefits of motor control, many conveniences of life that we enjoy and take for granted would also be denied us. Our environment is dominated by ac electric power. Wherever we are we can "plug in" to electric power. Yet because of the poor performance that has plagued the speed control of ac motors, we find many household as well as industrial applications using the dc motor. The thyristor, acting as a *controllable* rectifier, SCR, has many merits that have made it popular for motor speed control. We may not soon see the SCR supplanted by the power FET in some applications because of cost. Small appliances, hand tools, and fractional-horsepower dc motors that run off the ac mains may continue to use thyristors not merely for the economy but also because such control systems are simple. Nothing is simpler than using an SCR between the ac mains and a fractional-horsepower dc motor, as we can see in Fig. 6-2. Here we have an SCR doing double duty, acting first as a rectifier and second, acting as a very efficient rheostat by virtue of the variable firing angle achieved by the potentiometer. The resistor bridging from the armature to the SCR gate maintains constant speed irrespective of load variation. If we load the motor, causing it to slow down, the counter EMF generated within the armature winding will also drop and the gate potential will cause the SCR to fire earlier, which, in turn, will increase the armature voltage, pulling up the speed to its former value.

We may by this time have good reason to question if power FETs have any place in the control of dc motors in light of what we have just studied. So far our focus has been on fractional dc motors controlled from ac mains, and for these our answer may be negative simply because of cost.

A dc motor of greater horsepower that we would find serving in industrial applications would also use thyristors for speed control. To achieve greater horsepower it is not uncommon for us to find these motors operating from multiphase ac lines and at elevated voltages, generally running from 220 to 440 V. The reason is simple. We know that an SCR operating off commercial ac mains commutates by virtue of the alternating current. A 60-Hz line changes polarity every 8.33 ms, which is more than adequate for commutation of the SCR. We find multiphase lines to be helpful because we need more and steadier armature current to maintain the high torque for sustained horsepower. In

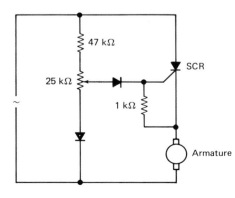

Figure 6-2 Half-wave SCR drive of a fractional-horsepower dc motor. (Timothy J. Maloney, *Industrial Solid State Electronics: Devices and Systems,* © 1979, p. 517. Reprinted by permission of Prentice Hall, Inc., Englewood Cliffs, N.J. 07632.)

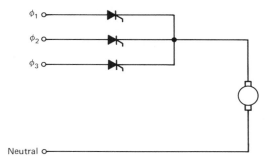

Figure 6-3 Basic three-phase drive using SCRs. (Timothy J. Maloney, *Industrial Solid State Electronics: Devices and Systems*, © 1979, p. 523. Reprinted by permission of Prentice-Hall, Inc., Englewood Cliffs, N.J. 07632.)

Fig. 6–3 we have a simple three-phase wye-connected circuit powering a dc motor using three SCRs. If we are clever in our gating of these SCRs we will be able to improve our armature current by nearly twice that of a single-phase system.

In earlier chapters we have been exposed enough to the advantages of power FETs over bipolar transistors, Darlingtons, and SCRs so as not to need another review. It should suffice for us to remember that SCRs commutate solely by the alternating nature of the incoming voltage. Power FETs do not need external support for commutation. When we apply a gate potential to our power MOSFET, we turn it ON, and when we remove the gate potential, the power MOSFET turns OFF. It would seem reasonable to us that if we choose to pass a higher armature current, we would only need to switch the power FET faster.

However, power MOSFETs present an interesting challenge to the prospective user who might wish to replace SCRs in an ac control system driving dc motors. In Chap. 4 we discussed in some detail what we then considered a beneficial parasitic, the source–drain reverse *p-n* diode intrinsic to the basic vertical structured MOSFET. If we chose to replace our SCR with a power MOSFET, we would soon discover that we have a unidirectional switch! As long as the applied drain voltage remains positive, our gate would control the current flow. However, once the drain potential reverses polarity, our power MOSFET conducts irrespective of gate voltage. The result is that our controller resembles a half-wave rectifier rather than a motor controller.

So our conclusions become a bit clearer. Power MOSFETs are excellent in motor controllers, but special design precautions must be taken when the mains are alternating current. In Chap. 9 we discuss how we can use power MOSFETs as analog switches (an ac line is an *analog voltage*). There are many applications for motor control where we will find the power MOSFET the only practical solution and we focus our attention on those applications in this chapter.

We must be sure that efficiency and practicality are best served if we chose power FETs. If thyristors do a better job at less expense, then by all means we should stick with the SCR. On the other hand, if we can become convinced as we finish this chapter that power MOSFETs offer advantages, we should endeavor to use them.

Dc motor control from direct current lines is quite a different story. The singular advantage of the SCR disappears: commutation from the alternating nature of the ac lines. To commutate an SCR in a dc circuit becomes a messy complication. Futhermore, we no longer need rectification: we already have a dc supply for a dc motor.

We might raise the question: Where would we have need of such a controller; what would be our application? Undoubtedly, the single largest industry in the United States is the automotive industry. In this industry everything eventually related to electrical control narrows down to the 12-V dc supply voltage offered by the storage battery. New automobiles are becoming electronic wonders and the microprocessor is taking charge over every task, such as controlling the *electric* fan used to cool the radiator and, of course, the electric window wipers, the electric window controls, and the electric fuel pump. In some of these situations the electric motor needs speed control, and in other functions we only need to turn the motors ON or OFF. Because of the ease in which we can drive power MOSFETs direct from logic (which we shall cover in detail in Chap. 8) and because of the unnecessary complication needed to use SCRs, power FETs appear to be a reasonable adjunct to the modern electronics in our automobiles. A simple circuit using a D/A converter and a quad op amp in a microprocessor-controlled dc motor control is shown in Fig. 6–4. In this circuit a low $R_{DS(on)}$ power MOSFET is used for two reasons. Motor controllers are subjected to large in-rush currents as the motor first begins. As the armature speed builds, the counter EMF reduces the armature current to a fraction of the startup current. A high $R_{DS(on)}$ would simply result in high losses and excessive temperature rise. A temperature rise would cause the $R_{DS(on)}$ to rise and if for any reason the motor stalled, a thermal runaway condition might occur, with possibly disastrous results. Furthermore, a low $R_{DS(on)}$ is desirable simply to reduce the effective V_{SAT} ($I_D R_{DS(on)}$). In this circuit the D/A sets the speed by establishing a reference voltage. Op-amp comparators a and b generate a triangle wave, which forms the basis for a PWM modulator resulting from op-amp comparator c, which, in turn, biases the MOSFET gate. The counter EMF helps establish speed control.

Another application where we would have dc mains and a micropower dc motor is the control of model railroads, where very sophisti-

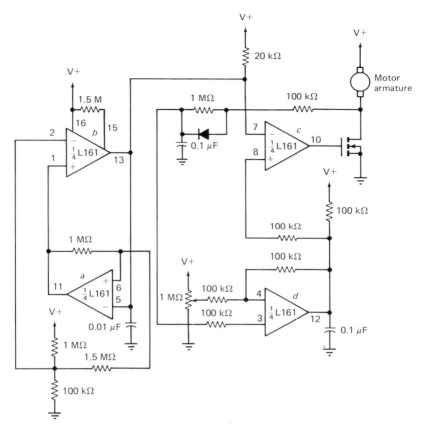

Figure 6-4 Constant-speed dc motor controller using power MOSFETs. (Courtesy of Siliconix incorporated.)

cated motor controllers already exist. Presently, these model railroad controllers use Darlingtons and power bipolar transistors and rely heavily on a form of *pulse-width modulation* (PWM), to effect what is called by these hobbyists as prototype railroad movement. An interesting function also popular among model railroad enthusiasts that we would find especially suited for power MOSFETs is called *momentum*, or what amounts to inertia simulation. All we do to effect momentum is to charge a capacitor which biases the MOSFET gate so that we neither force an instant start nor an instant stop. This is very easy with power MOSFETs because of their inherently high gate resistance and quite a bit more complicated with either power bipolar transistors or Darlingtons.

Possibly one of the easiest dc motor controllers is to use power FETs to control the movement of a brushless stepping motor. What becomes especially attractive is how easy it is for us to interface this stepping motor directly to computer logic. Stepping motors find wide use in many computer-related applications, and the ease of interfacing direct to a computer through the use of power MOSFETs is highly welcomed by designers. The many advantages inherent in power FETs, such as their attractive SOA features and especially the parasitic source-drain *p-n* diode, allow us the luxury of a simple drive circuit, as shown in Fig. 6-5, where we see direct logic drive and no free-wheeling diodes for snubbing networks.

Our final illustration where we would have a dc motor controlled from dc mains is a traction-drive vehicle, in other words, an electric vehicle. A dc motor offers incredible starting torque, which is highly desirable, but as we know, torque is proportional to armature current, so we would not wish to control the speed using series resistors, as efficiency would be greatly impaired. Generally, for such service we would not use SCRs because of the difficulty of commutation but, rather, germanium power bipolar transistors as series switch elements. A germanium transistor offers a lower V_{SAT} than a comparably rated silicon transistor. PWM would provide smooth and quiet motor opera-

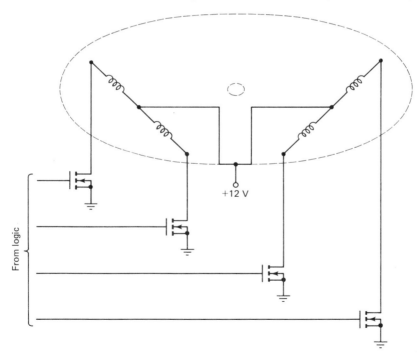

Figure 6-5 Direct logic-control stepping motor drive with power MOSFETs. (Courtesy of Siliconix incorporated.)

tion.   Under high-torque situations, such as climbing a grade, PWM would provide high armature current and low armature voltage for optimum performance.   A danger, of course, would be that we might exceed the safe operating area (SOA) of these series pass transistors.

Of course, power MOSFETs have not yet reached the stage of development where they could be substituted for high-current bipolar transistors.   We saw earlier that $R_{DS(on)}$ is proportional to breakdown voltage [see Eq. (4.13)] so it is conceivable that an ultra-low $R_{DS(on)}$ power MOSFET could be fabricated that would operate directly from either 12- or 36-V batteries for vehicular traction.

In these few pages we have hardly covered the many types of dc motor controllers that exist today.   We have found that thyristors offer excellent performance in controllers fed from ac mains, where easy commutation of SCRs allows both simple and reliable circuits.   Although dc motor controllers are both simpler in design and obviously very popular, nevertheless the dc motor has some inherent faults that makes the ac induction motor an attractive alternative to equipment designers.   Notable among the faults of the dc motor is the interlocking behavior between speed and torque.   Whenever the load is relaxed, a dc motor will speed up, and conversely, whenever the load is increased, the motor will slow.   In some ways this interrelationship does provide a modicum of speed control.   Complicating this interrelationship, however, is the effect that speed plays on armature current and the counter EMF.   In Fig. 6-4 we viewed a simple dc controller with feedback for motor speed control.   The effectiveness of this simple EMF feedback lies in the constant-current characteristics of the power MOSFET.   For the most part, speed controllers for dc motors rely on tachometers for accurate feedback in SCR systems.   In dc motor control from ac mains we can appreciate that the power MOSFET would be hard-pressed to replace the thyristor.   SCRs are capable of handling extremely high currents and withstanding very high voltages, far beyond anything yet possible with power MOSFETs.   The high inrush currents typical of the dc motor when first starting would tax the performance of the power MOSFET to the point where the SOA might be in jeopardy.   In dc motor control from dc mains, or, as we saw in Fig. 6-5, from logic control, the power MOSFET stands out as a superior performer worthy of careful consideration by any designer.

6.3 AC MOTOR CONTROL

Equation (6.1) identified the means by which we can attain speed control for the ac induction motor—adjustable frequency control.   A major advantage of the ac induction motor that makes it highly desirable for many applications is that its speed is not interlocked with torque to the extent that we previously saw for the dc motor. If we were to load an induction motor, its speed would remain reasonable constant, to the point where our loading would cause excessive slippage, at which time we would witness a stall. An ac induction motor does not rotate at the synchronous speed but actually slips behind the rotating field by a small amount, possibly by 5 to 10%. We can compare the speed-torque relationship of the dc motor with that of the ac motor in Fig. 6-6, which dramatically illustrates the advantage of the dc motor at low speeds, and the advantages of the ac induction motor under variable load.

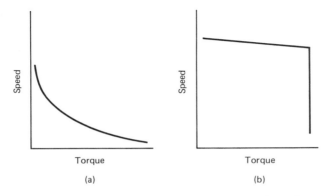

Figure 6-6 Relative comparison of speed vs. torque for a dc motor (a) and an ac motor (b).

There are, however, other mitigating factors that have led to the use of the ac induction motor that we need to be made aware of. In 1976, a U.S. government report revealed some very surprising statistics when it identified the fact that electric motors alone consume nearly two-thirds of the total electricity used in the United States! We need not dwell on the need for ac motor speed control as a source of energy conservation. We are aware of many applications where we would want variable speed control. Many household as well as industrial needs could be better served with a variable-speed ac motor rather than one running at full speed with some mechanical or fluid clutch providing speed control. One example might be the home air conditioner; another example might be an industrial pumping station.

The typical ac induction motor drive is shown in Fig. 6-7, where we have what amounts to an *inverter*. An inverter changes dc power into a variable frequency suitable for driving the ac induction motor, and more often than not the means of obtaining this variable frequency

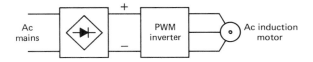

Figure 6-7   Block diagram of a basic ac induction motor drive.

ac power is by PWM.   Pulse-width modulation is a highly desirable method used to drive ac induction motors, as it offers great flexibility of output waveform.   Driving an ac induction motor requires that we maintain a low harmonic waveform, and as the speed is varied, we must also vary the ac voltage to the motor so as to prevent saturation of the field.   PWM provides both of these requirements at minimum cost and maximum reliability.   We are able to effect a constant voltage-to-frequency ratio which optimally can control an ac induction motor.

Historically, ac induction motor speed control followed a pattern similar to dc motor speed control: both relied heavily upon the SCR. In multiphase ac induction motors these SCR controlled-speed controllers can often cost considerably more than the motor under control and because of the tricky gating required to provide the multiphase ac power, the timing was both complex and often not overly reliable. Besides these shortcomings, the SCR does not lend itself to providing good sinusoidal waveforms, and as a consequence the ac induction motors were not able to run at peak efficiency.   Overall we can understand the popularity of the more costly and less reliable dc motor. Despite its shortcomings it was often cheaper to use in the long run. But that was "yesterday."   Today we are in a new world, where energy costs are a major lever in our economy and we must seek every avenue open to us that will lead to conservation.   The ac induction motor offers us economy, longer life, less maintainability, and more reliability than does the dc motor. Since the invention of the ac induction motor we have been locked into the 50/60-Hz syndrome and we can closely approximate the horsepower of a motor by its size and weight.   A 25-hp ac induction motor weighs about 400 lb.   But we could design an ac induction motor to operate at higher frequency and, as a result, our motor would be smaller, lighter, and possibly more efficient than its predecessor.   Military aircraft have used high-frequency ac induction motors for years, fitting them into areas too cramped for 60-Hz motors. That 25-hp ac induction motor designed to operate at 400-Hz would weigh about 30 lb!   Equation (6.1) identified our speed equation; an increase in frequency means more poles for equal speed. PWM offers us the easiest route to effective speed control for ac induction motors and the challenge to develop high-frequency, cost-effective ac induction motors.

6.3.1 Drive Modulation

Numerous pulse-width modulation schemes have been reported over the years that we will not try to discuss within the limits of this book. Rather we focus our attention on two types of drive modulation. The first type of drive modulation that we shall examine we call stair-step modulation. This form of modulation has been very popular with thyristor control, where it was also called *adjustable voltage input*, (AVI) drive. Stair-step or AVI drive is best understood by illustration, as shown in Fig. 6-8. Since the greater majority of ac induction motor

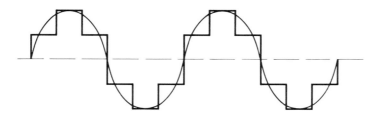

Figure 6-8 Adjustable voltage input (AVI) or stair-step drive waveform.

speed controllers are for three-phase motors, we generally refer to the stair-step as a *six-step drive* simply because the implementation of the waveform results from the precise sequential switching of six thyristor switches arranged as shown in Fig. 6-9. It should be obvious that SCRs switch either full ON or full OFF, with, of course, the option of adjusting the firing angle. The waveform that we saw in Fig. 6–8 results from driving each totem-pole pair of thyristors with a square wave, each leg being $60°$ out of phase. A closer examination of the switching wave-

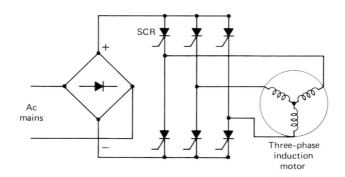

Figure 6-9 Three-phase induction motor control using SCRs.

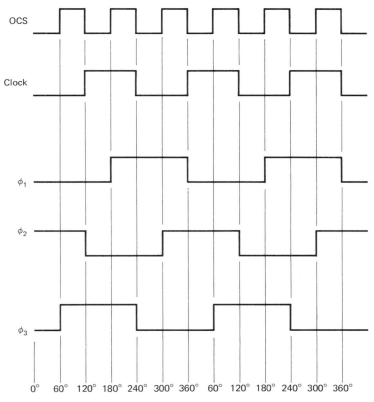

Figure 6-10 Development of the switching waveform to drive a three-phase motor.

forms is offered in Fig. 6–10, where their relationship to the clock frequency is evident. The resulting sine wave is caused by the smoothing field inductance of the induction motor.

The firing angle of the thyristors plays an improtant role in this six-step motor drive. Since the output voltage is directly proportional to the dc mains, we are able to fix the output voltage by the control of the firing angle. We earlier noted the necessity of voltage control as the frequency is changed. In this AVI modulation, combining the logic to implement the stair-step and simultaneously adjusting the firing angle is an additional complication that has helped to popularize pulse-width modulation for multiphase ac induction motor drives.

A novel stair-step motor drive which combines a form of PWM for torque control is shown in Fig. 6–11. The entire drive is assembled from easily obtainable circuits and the motor capacity (in horsepower)

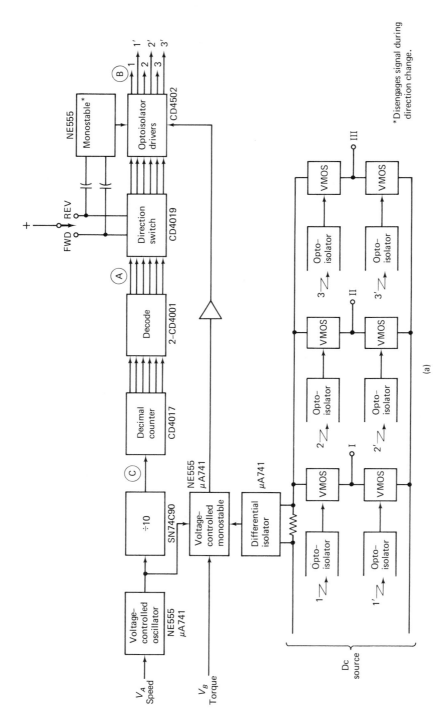

Figure 6-11 (a) Block diagram of a PWM three-phase motor control showing waveforms (b).

170

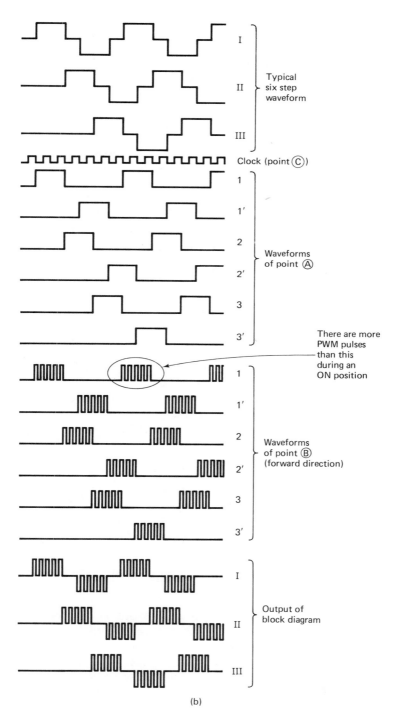

I

II — Typical six step waveform

III

Clock (point C)

1

1'

2

2' — Waveforms of point Ⓐ

3

3'

There are more PWM pulses than this during an ON position

1

1'

2

2' — Waveforms of point Ⓑ (forward direction)

3

3'

I

II — Output of block diagram

III

(b)

Figure 6-11 Continued.

is determined by the voltage and current rating of the power MOSFETs in the totem poles. A major shortcoming of this and other ac induction motor drives is the need to use optoisolators between the motor and the main control. This is a requirement of Underwriters' Laboratories, Inc. Directional control simply reverses phases 1 and 3, and the monostable momentarily disconnects the drive during directional control switching. The differential isolator monitors the average current of the motor and drive and combined with the pulse timing circuit and the op amp within the voltage-controlled monostable provides a variable-pulse-width gating signal that, once inverted, directly gates the optoisolator drivers. Noticeably absent in this power MOSFET drive, as well as in all power MOSFET drives, are the reverse-current surge-protection diodes. These diodes are, as we well remember from our discussion in Chap. 4, *beneficial parasitics* that can easily handle as much current as the MOSFET itself. Their absence in a thyristor motor control could result in false triggering and in a bipolar transistor controller might result in catastrophic second breakdown.

The second type of drive modulation that we will explore is *pulse-width modulation* (PWM). We are not only freed of the bothersome independent control of output voltage as the frequency (or speed) varies, but the harmonic content of the waveform is greatly reduced from what we would have were we to use either AVI or the more conventional thyristor control. We should always keep in mind that harmonics represent energy that is unusable by the ac induction motor. Consequently, as we lower the harmonic content of our drive, the overall efficiency improves.

We should perhaps quickly review the importance of voltage control in ac induction motor speed controllers. An ac motor consists of both resistance and inductance. If we vary the frequency to achieve speed control [Eq. (6.1)], we also vary the *inductive reactance* of the motor, and this, in turn, varies the current through the motor. As frequency varies, the dc resistance remains unchanged, but this inductive reactance, X_L, varies proportionally:

$$X_L = 2\pi f L \qquad\qquad (6.2)$$

As we change our motor speed by changing frequency [Eq. (6.1)] we must change the impressed voltage across our ac induction motor to maintain a constant current. This constant field current establishes a fixed rotating magnetic field which maintains a constant torque. Our controller, to maintain constant torque, must provide a constant V/Hz ratio, as shown in Fig. 6-12. If we failed to control voltage as our speed changed, the torque would also change. Torque varies as the square of the field strength, and as the speed reduced because of a lower frequency, the field current would increase. This conceivably

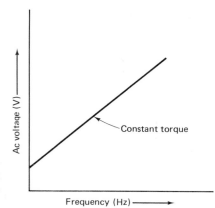

Figure 6-12 Voltage-frequency relationship as frequency is adjusted to effect a speed change while maintaining constant torque.

would result in, first, increased torque; second, field saturation; and third, possible damage to the motor.

By properly sequencing a pair of power semiconductors arranged in a totem pole, we can develop PWM control of an ac induction motor. An overly simplistic illustration is shown in Fig. 6-13. The amplitude of the sine wave is proportional to the width and phase of the pulse. PWM can consist of either a fixed ratio, where the number of pulses per half-cycle remains fixed and only the width varies, or we can have a variable ratio, where the number of pulses increase with the pulse

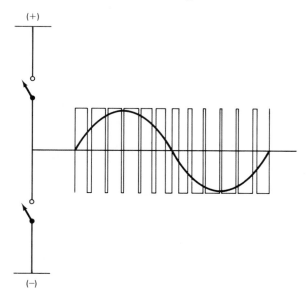

Figure 6-13 PWM waveforms can be generated from a simple totem-pole switch pair.

period and the width varies for voltage control. There are, of course, various embellishments of these basic types that can be found in many industrial sites.

The heart of PWM is the generator itself. We can easily synthesize a PWM waveform using an op amp as our basic building block. In Fig. 6-14 we have an op amp such as a 741 or an LM356 fed with a sawtooth waveform into its inverting input (-) and a dc voltage impressed upon its noninverting input (+). If we vary the clock frequency, which, in turn, changes the sawtooth frequency, we are able to change the number of pulses per second, and by changing the dc voltage level we can change the pulse width.

The ac motor control that we have been studying relies first, upon conversion of the ac mains to rectified direct current, which then passes through switching semiconductors to simulate ac, from which we energize the motor. Our switching modulator performs the task of transforming the rectified dc into as nearly sinusoidal a waveform as possible. If our motor is multiphase, then, of course, our modulator must perform the necessary phase shifts to feed each leg in its proper phase. To accomplish this task regardless of semiconductor, we need to use the totem-pole arrangement, alternatingly—not simultaneously—switching first the upper and then, in sequence, the lower. We might easily visualize these as simple switches. As we switch alternatingly between the upper totem-pole element and the lower we, in effect, cause the dc voltage to alternate in polarity in like fashion. However, if

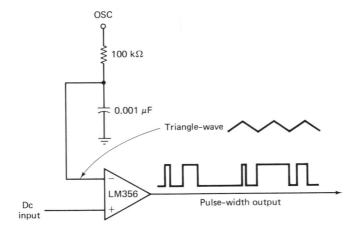

Figure 6-14 PWM waveform generator results from the merging of a triangular wave, which controls the PRF (pulse repetition frequency), with a dc level, which changes the pulse width.

this was all we did, our ac output would not be sinusoidal but a square wave. It is because of this that we have considered two forms of modulation, the stair-step (AVI) and pulse-width modulation (PWM). Both of these forms of modulation allow us to simulate sinusoidal waveforms more closely. Were we to allow square-wave modulation, we would be plagued with rich harmonics that, as we have said, could not be utilized by our ac motor other than to cause excessive heating and reduced efficiency. Aside from our careful design of a modulator, we must now consider the semiconductor switching elements.

6.3.2 The Totem-Pole Switch

The thyristor has been the favorite control switching element for dc motors driven from ac mains for reasons discussed in some detail earlier in this chapter. Driving ac motors from dc mains, however, has not proven so easy. The fundamental problem with the thyristor has been in commutation, and the overall complexity has proven to be such a burden that, for the most part, it has seen little use. Low-horsepower multiphase ac induction motors have, with some success, relied on speed controllers using Darlingtons as the totem-pole switch elements; other controllers have relied on germanium power bipolar transistors. In either case our motor controllers have been unduly complicated because of the drive requirements of these totem-pole semiconductor switches, which has necessitated the use of isolated driver amplifiers and floating power supplies. A block diagram of a typical system is shown in Fig. 6–15.

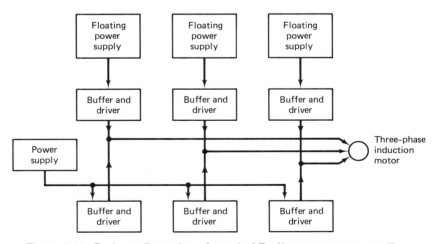

Figure 6-15 Basic configuration of a typical Darlington motor controller.

In this chapter we have addressed the control of two quite different motors, the dc motor, which because of thyristors has become the more popular, and the ac induction motor, which because of its high reliability and lower cost has been the dream of designers everywhere but whose control has been a nightmare.

We may probably never see power FETs in any form either match or surpass the incredible power-handling capability of the thyristor. For this reason large industrial motor applications will probably use thyristors for many years to come. We should not be too surprised someday to discover power FETs operating at or beyond 1000 V with current-handling capacity of several tens of amperes if technology can resolve the associated $R_{DS(on)}$ problem [see Eq. (4.13)]. If their SOA follows, power FET motor controllers will find a wide-open market in which ac induction motors will supplant dc motors in areas such as rapid transit vehicles and other critical areas where several motors are required to run synchronously for optimum performance.

Unlike the benefits we needed for high-speed switchers and regulators described in Chap. 5, the power semiconductor switches for motor control do not need high-speed switching capability. The lack of minority-carrier storage time common to all FETs is not a major benefit in motor controllers. Why, we might ask, would we use power FETs in motor controllers?

Economically, we probably would prefer thyristors in controlling small and fractional-horsepower dc motors operating from ac mains. Nothing appears as easy and low cost as having an SCR feeding a household appliance or a small hand tool. However, we have a new problem when we try controlling an ac motor or even a dc motor from dc mains. Here our attraction to the power FET focuses on several quite fundamental advantages. Because this unique transistor does not have minority-carrier injection, it exhibits a high input resistance which greatly simplifies the driving requirements. Very little input power will control large currents. Simple logic drives suffice in many situations. Totem-pole arrangements are easily driven where for a Darlington or power bipolar transistor we would need complicated drive circuitry. We can also eliminate buffering between stages, which further simplifies the complexity of our drive circuits. A major impact that power FETs also share with Darlingtons and power bipolar transistors is obvious: the easy commutation. When the drive is removed, the semiconductor shuts down irrespective of either load conditions or the polarity of the ac mains. If we were ever to plot the behavior of a highly reactive motor on the load line, we would be especially grateful for the much improved second breakdown characteristic of power FETs as compared to power bipolar transistors. A plot of a highly reactive motor load line is shown in Fig. 6-16 superimposed over a "typical" transistor's SOA.

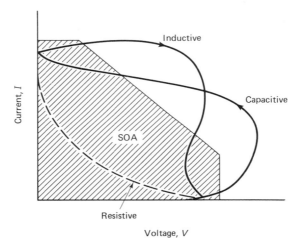

Figure 6-16 Highly reactive loads caused by motors can easily exceed the safe operating area (SOA) of power transistors.

Another troublesome area when using power bipolar transistors in motor control applications results under high collector currents. Invariably, the bipolar transistor's gain h_{FE} droops badly. The properly constructed power MOSFET suffers no such problem. We can compare h_{FE} and g_m of a comparably rated bipolar transistor and power MOSFET in Fig. 6-17.

Undoubtedly, the singular shortcoming of the power FET is, as before, its dependency of breakdown voltage to $R_{DS(on)}$. Acting as a switch element in motor controllers we are not so concerned with the saturated region but we *are* concerned with simple V_{SAT}-type losses. We can, of course, parallel our power MOSFETs to ensure lower $R_{DS(on)}$ and higher operating current with no loss in drive requirements. For the most part we are unconcerned with input capacity, as our switching speeds are miniscule and hence our charging currents are also insignificant.

As power FETs increase in their breakdown and current handling, we shall see increasing emphasis on ac three-phase motor controllers replacing costly dc motors in every area.

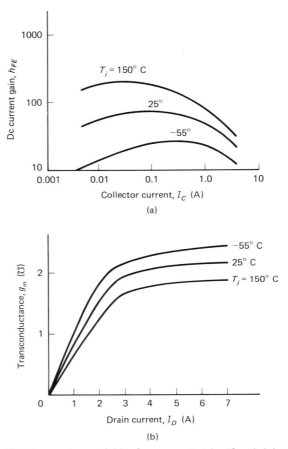

Figure 6-17 Comparison of bipolar current gain (h_{FE}) (a) to the forward transconductance of a power MOSFET (b).

REFERENCES

BENDANIEL, D. J., and E. E. DAVID, JR., "Semiconductor Alternating Current Motor Drives and Energy Conservation," *Science*, 206, No. 4420 (Nov. 16, 1979), 773–76.

CLEMENTE, S., and B. PELLEY, "A Chopper for Motor Speed Control Using Parallel Connected Power HEXFETs™," *International Rectifier Application Note 941*, El Segundo, Calif., International Rectifier Semiconductor Division, 1981.

EVANS, ARTHUR D., et al., "Higher Power Ratings Extend VMOS FETs' Dominion," *Electronics*, 51 (June 27, 1978), 105–12.

FAY, G. V., and NICK FREYLING, "Pulse-Width Modulation for D-C Motor Speed Control," *Motorola Application Note AN445*. Phoenix, Ariz.: Motorola Semiconductor Products Inc., 1972.

GUTZWILLER, F. W.,"Thyristor and Rectifier Diodes—The Semiconductor Work-horses," *IEEE Spectrum*, 4, No. 8 (Aug. 1967), 102-11.

HARDEN, JOHN D., and FORREST B. GOLDEN, eds., *Power Semiconductor Applications*, Vol. I: *General Considerations*. New York: IEEE Press, 1972.

JOOS, GEZA, and THOMAS H. BARTON, "Four Quadrant DC Variable-Speed Drives—Design Considerations," *Proc. IEEE*, 63, No. 12 (Dec. 1975), 1660-68.

KUSKO, ALEXANDER, "Solid-State Motor-Speed Control," *IEEE Spectrum*, 9 No. 10 (Oct. 1972), 50-55.

MALONEY, TIMOTHY J., *Industrial Solid-State Electronics*. Englewood Cliffs, N.J.: Prentice-Hall, Inc., 1979.

PRESSMAN, ABRAHAM I., *Switching and Linear Power Supply, Power Converter Design*. Rochelle Park, N.J.: Hayden Book Company, Inc., 1977.

WEBER, HOWARD F., "Solid-State DC Motor Control for Traction Drive Vehicles," *Motorola Application Note AN189*, Phoenix, Ariz.: Motorola Semiconductor Products Inc., 1974.

seven

Using the Power FET
in Audio Power Amplifiers

7.1 INTRODUCTION

Ever since about 1962, when the bipolar transistor began in earnest to replace the vacuum tube in audio equipment, the critical audiophile has argued that what he now hears is somehow "different." Perhaps for that reason we still find many of the top-of-the-line high-fidelity audio power amplifiers built with vacuum tubes. Controversy continues to swirl within the high-fidelity camps, arguing whether the vacuum-tube amplifiers "sound better" than do transistor amplifiers. Their language is equally colorful, with such expressions as *boom*, *warm*, *brittle*, and *metallic*. Among these experts who feel certain that they, indeed, hear something different, we find that, more often than not, they have aligned themselves within the vacuum-tube camp, and for justification they cite, with obvious bias, the benefits of vacuum tubes and the faults of the bipolar transistor. We shall try a more balanced approach.

There are many high-fidelity power amplifiers available today. Everywhere we turn we find all solid-state equipment, and, of course, some vacuum-tube equipment. Phonograph and tape decks are popular, as are stereo FM and now possibly stereo AM radios, offering the listener high-quality sound through all solid-state equipment. Admittedly, solid-state offered us immediate advantages over vacuum tubes that won the affection of many. Portability was certainly a major advantage, not simply because of the weight savings but because batteries could replace the inconvenience of plugging into a power line. Perhaps the finer distinctions now criticized by audiophiles were masked by small loudspeakers that helped make the equipment portable. The transistor offered major benefits, such as greater reliability, lower power consumption, miniaturization, and certainly not to be overlooked, they made possible the elimination of both interstage and output transformers and a moderate reduction in the size and complexity of power transformers. So not only did we have portability from the vantage point of not being dependent upon commercial ac mains, but we were relieved of much of the dead weight formally attributed to the transformers. As our all solid-state equipment matured, we benefited financially, finding greater flexibility in both usage and features in our all solid-state equipment. But did we sacrifice quality of reproduction? Is what we hear "concert quality?" And do we care if it is or not? Are the only critics of the bipolar transistor audio-amplifier audiophiles, or can we all detect a "difference?" Let us begin by comparing the features of power FETs with those of the bipolar transistor *and* the vacuum tube.

7.2 FETS, TUBES, AND BIPOLAR TRANSISTORS

We did not leave much uncovered in Chap. 4 pertaining to power FETs that we cannot find when we begin to compare the power FET with either the vacuum tube or the bipolar transistor. If our memory serves us, we may recall that the high-fidelity audio amplifier of a generation ago was usually built around the power triode vacuum tube or the triode-connected pentode, and that for good reason. We begin our study of comparisons by examining the features of the triode vacuum tube. When we are through, we will have gained a better appreciation for its nickname, *thermionic FET*.

The output characteristics of a triode, shown in Fig. 7–1, offers us several benefits that makes it ideal as an audio output stage. Because of its high output conductance, g_{os}, it more closely matches the speaker impedance than would a low-conductance (high-output-resistance) bipolar transistor. Again, because of its triode characteristic—which is what we generally call the curves of Fig. 7–1—a triode offers lower high-order odd-harmonic distortion in a push-pull output circuit. Now if we superimposed a reactive load line over these output characteristics, we would understand the reason for lower distortion. In Fig. 7–2(a), we have superimposed a highly inductive speaker load line which has taken the shape of an elongated elipse. A duplicate load line superim-

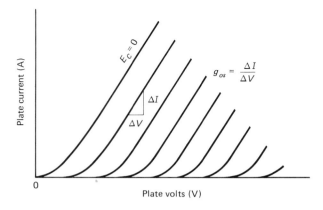

Figure 7-1 Output characteristics of a typical triode vacuum tube.

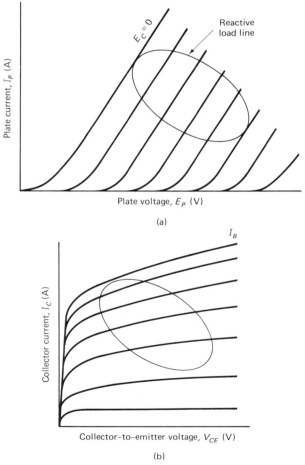

Figure 7-2 Nonlinear bipolar transistor's output characteristics within the area encompassed by the reactive load line, showing noticeable cause for distortion when compared with that area of the triode vacuum tube.

posed over a typical bipolar transistor's output characteristics is shown in Fig. 7-2(b). The cause for distortion in the bipolar transistor is easily evident, whereas we find little cause for distortion in the triode. Super-imposing an eliptical load line only tells us part of the reason for low distortion; a fuller treatment requires us to remember the tranfer equations for both the vacuum tube and the bipolar transistor [see Eqs. (4.1) and (4.4)]. Where the triode vacuum-tube plate current behaves according to the three-halves power, the collector current of the bipolar transistor is highly nonlinear. Because of this nonlinear transfer characteristic we can easily visualize the generation of harmon-ics, hence distortion products, that would occur in a bipolar transistor stage. Part of our success in achieving low distortion in a triode vacuum tube, however, lies in having a high idling current to avoid the curvature of the static characteristics found at low plate current.

A very major advantage of the power vacuum tube in a push-pull output stage is that we experience no storage time and no turn-ON delay when we apply gate voltage. We are familiar with the bipolar transistor's storage-time problems and also with the base–emitter threshold voltage, which tends to cause some delay in turn-ON. One unique advantage offered us by the vacuum tube is that it will forgive us for exceeding its SOA. The amount of abuse that a vacuum tube can take is truly remarkable.

Many of the advantages that we have covered for the vacuum-tube triode also exist for the vertical J-FET. Unfortunately, all transistors, FETs included, have a SOA. The static induction transistor (SIT) is basically a power J-FET with triodelike characteristics when the gate is biased *negatively*. Consequently, many of the advantages that we have just reviewed are quiet applicable to the SIT. Like the vacuum tube, we find that we must operate at a high quiescent drain current for optimum performance. Actually, we could obtain comparative distor-tion performance using bipolar transistors if we raised their collector current to similar levels, but we would have a major thermal problem that could be disastrous if not kept under tight control. We generally associate FETs with degenerate thermal properties; in other words, as the temperature increases, we *expect* to see the drain current decrease. A serious disadvantage of the SIT is that we must bias the gate nega-tively because the SIT, like all J-FETs, is a depletion-mode device. If, for any reason, we allowed the gate-to-source voltage to return to zero volts, maximum drain current would flow and problems could quickly develop. If, for example, we were using a pair of complementary SITs, one *n*-channel and one *p*-channel in a typical push-pull, totem-pole arrangement and lost our gate bias, we would have a short-circuit across the dc mains. This poses a problem for the designer of a SIT amplifier since bias voltage *must* be applied before the drain supply is energized.

FETs and bipolar transistors have one very major advantage over the vacuum tube: they can be complementary. That is, we can have an "upside-down" transistor: in other words, a p-channel FET or a pnp bipolar transistor. It works exactly like its counterpart, the n-channel or npn, except that all voltages are reverse polarity. The only real shortcoming to this p-channel complementary transistor is that we find that its carrier mobility is considerably less than the n-channel transistor. To effect a perfect amplifier, we need matched stages. If, for example, the p-channel SIT has one-third less mobility than the n-channel, we need a device one-third larger to match the n-channel's performance. So we need to qualify what we mean when we say that we have a complementary pair. What parameters match? For the SIT we have three boundaries: the pinch-off voltage V_p, the forward transconductance g_m, and the zero-bias drain current I_{DSS}. A matched complementary pair would present a transfer characteristic similar to that shown in Fig. 7-3. In a high-quality, high-fidelity audio amplifier we would have to provide a good match between the n- and p-channel complementary pair if we wanted to achieve high fidelity with low distortion. Since to achieve complementary pairs we have a p-channel FET larger than its n-channel partner, we then recognize that some features are obviously not complementary, the most notable being capacity. Since we have a physically larger p-channel FET we obviously have larger capacitance. A larger input capacitance (including the Miller effect) requires that we

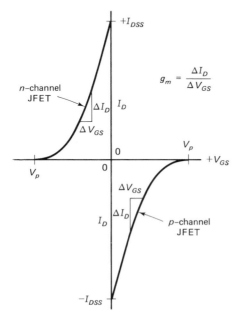

Figure 7-3 Transfer characteristics of complementary static induction transistors.

drive the p-channel FET harder to overcome the $C\, \partial V/\partial t$ and RC time constant.   For this reason we may find that for many situations the quasi-complementary push-pull audio amplifier is often more suitable for our needs where we would use only n-channel FETs.

Much of what we have covered to this point has focused upon the SIT and if the SIT has shown itself as such a splendid candidate for use in high-fidelity audio amplifiers, we might ask why it has not been more widely used.   There were a couple of reasons.   One problem was $R_{DS(on)}$, which, coupled with the heavy drain current required to pump out appreciable audio power, simply caused excessive dissipation and, as we might expect, poor efficiency.   A high $R_{DS(on)}$ (typically 5 Ω) coupled with a high quiescent standby drain current made the SITs run very hot.   Another potential problem, which we touched upon in Chap. 2 (see Fig. 2–18), is that when the gate-to-source voltage of the SIT swung *positive*, the typical depletion-mode J-FET characteristic looked remarkably like that of a bipolar transistor!   If during heavy usage with the SITs running hot, we had a crescendo we could overdrive the SITs and before the music died, our SITs would have entered into second breakdown, thermal runaway, and quite likely suffered catastrophic destruction.   Although all semiconductor power amplifiers need heat sinks, we must be sure to provide sufficient heat sinking for the SIT.

Another candidate for the output stage of a high-fidelity amplifier is the power MOSFET.   Although we have similar benefits to those of the SIT we have just discussed, the power MOSFET offers us additional benefits.   The power MOSFETs that we have singled out for study in this book are called enhancement-mode MOSFETs, which simply means that unless we bias them positively for the n-channel and negatively for the p-channel, no drain current will flow.   We have, in effect, fail-safe transistors.   If our bias should fail for any reason, we would not experience a runaway condition.   This was not so for the SIT, a depletion-mode FET.

Were we to compare the characteristics of an SIT with those of a power MOSFET, we would be impressed with the much lower imput capacity of the MOSFET.   A low input capacity (including the Miller contribution) is very desirable in improving the bandwidth response of the amplifier.   Figure 7–4 compares the bandwidth of a typical SIT with that of a 2-A power MOSFET.   Both are used in a simple Class A circuit.   The improvement that we obtain with the MOSFET is quite remarkable, showing an order-of-magnitude improvement in frequency response.

However, the power MOSFET is not without fault.   We read in Chap. 2 that the threshold voltage of a MOSFET is inversely proportional to temperature.   That is, as we raise the chip temperature, the

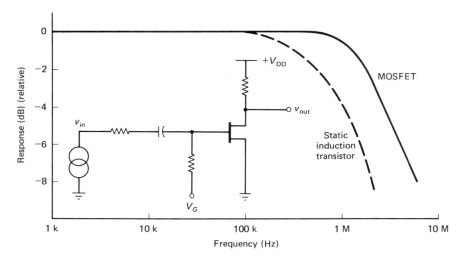

Figure 7-4 Comparative frequency response between the SIT and a short-channel MOSFET, showing the extended range available by virtue of the lower MOSFET capacity.

threshold voltage will drop. In fact, were we to raise the temperature too high, our enhancement-mode power MOSFET might revert to the depletion-mode. In our discussion of the SIT we learned that to achieve optimum performance we must bias the FET for a high quiescent drain current. High current suggests high temperatures. The problem we face is simple. As the operating temperature of our power MOSFET increases, its threshold voltage decreases. Since our bias is strictly dependent upon this threshold voltage level, we discover that we must track our biasing with threshold if we wish to maintain our amplifier operating in a particular class of service, such as, for example, having our amplifier in Class AB_1. Were we simply to ignore the temperature effects on threshold, we might begin with a cold amplifier biased to operate in Class AB_1 only to find that, once up to its operating temperature, the overload current relay opens, or, lacking a relay, if we are fortunate, a fuse blows. More than likely, however, we would burn out one or both the power MOSFETs. Why? Our Class AB_1 power amplifier is built either push-pull using complementary power FETs or quasi-complementary using a pair of n-channel power FETs. Either way, we would have a totem-pole arrangement across the dc mains. If either or both the power MOSFETs went into the depletion mode, the drain current could soar beyond the SOA. To the audiophile, a more subtle problem arises before this somewhat catastrophic problem: with a change in bias, the quality of reproduction shifts; the distortion levels shift as well as the open-loop gain.

Since the early 1960s the bipolar transistor has played an increasingly greater role in new audio-amplifier designs, and many critics will quickly add that its popularity is not because of improvements in fidelity. There is actually very little to commend the bipolar transistor for use in high-fidelity audio amplifiers. Its gross nonlinearities are overcome with large amounts of negative feedback, which, in turn, raises new problems with regard to *transient intermodulation distortion* (TID). The regenerative thermal properties of the bipolar transistor demand that we take special care by designing into our power amplifier shutdown circuitry to prevent catastrophic destruction resulting from either thermal runaway or second breakdown. And then, of course, we have the criticism from the audiophile, who quickly adds that a power bipolar transistor audio amplifier does not have the pleasing quality found in a vacuum-tube high-fidelity audio amplifier. One reason we find for this criticism results from the minority-carrier storage time of the power bipolar transistor, which causes what is known as *crossover distortion*, as well as a host of other related problems that in one way or another result in increased distortion. A somewhat exaggerated view of crossover distortion is shown in Fig. 7-5.

We might question why power bipolar transistors have taken such a strong leadership position in power amplifier design. The answer appears to be simply that bipolar transistors have been with us since 1950; we understand their workings better, we have dealt with their problems before, and perhaps we feel more "comfortable" using them, despite their problems. We are willing to accept problems that we know how to handle rather than accept new, and perhaps sinister problems that may still be in need of solution. This chapter endeavors to seek solutions that will show the advantages of using power FETs in high-fidelity audio amplifiers.

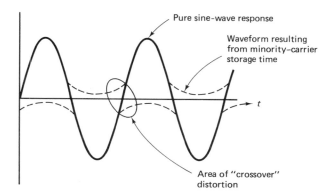

Figure 7-5 Crossover distortion inherent in a bipolar transistor push-pull amplifier caused by storage time.

We will not endeavor a detailed synthesis of audio amplifier design, as that is quite apart from the overall theme of the book. Rather, we will discuss certain important aspects that a designer should bear in mind and try to identify what features power FETs offer that will offer the most satisfactory solutions.

Our major goal is, of course, to achieve high fidelity that sounds like we are actually listening to the performers in real life as a live audience might hear them. We want no more distortion than that which originates in the recording studio. Any other goal that we might wish to achieve is subservient to our major goal of high-quality sound reproduction.

There are several forms of distortion, any of which will cause harm to the fidelity of our system. Distortion due to nonlinearities, distortion due to fast transients, and crossover distortion are among the major problems that most designers try to resolve. Distortion caused by overload, including that phenomenon called *sticking* (where, because of minority-carrier storage time, the collector current does not follow the base drive, leaving the bipolar transistor literally "stuck" to the power rail), perhaps requires a steadier hand on the volume control, although selecting the right transistor certainly helps to a degree. Of course, if overload were graceful, we would probably not object too strenuously.

To achieve low distortion, we generally find considerable negative feedback looping back from the output to an earlier stage. The more nonlinear the output stage, the more negative feedback we need. When a signal is fed back to an earlier stage and injected out of phase, degeneration occurs simply because this out-of-phase signal will cancel a portion of the incoming signal. The stronger this returned signal becomes, the greater is the gain reduction. We often associate negative feedback with gain-flattened amplifiers exhibiting extraordinary wide bandwidths. To ensure stability we also find phase compensation as part of the feedback network. Were we simply to plot the gain–bandwidth of our audio amplifier, we might be quite pleased with its performance. However, if we were a critical listener, we might condemn it, and for good reason. *Transient intermodulation distortion* (TID) is the reason for our disappointment. A sharp crescendo sends a series of fast pulses through our amplifier, so fast that our feedback network cannot respond quickly enough to provide the necessary degeneration. Instead, we find these fast pulses being returned to the earlier stages both out of phase and out of time with the incoming pulses, and the result is severe distortion. The cure? We must either reduce the amount of negative feedback or improve what is called the *openloop gain*. If we are using bipolar transistors, we may discover that neither task is easy, but with power FETs we have options open to us.

In many commercial high-fidelity audio amplifiers, one way we find to reduce various distortions caused by poor slew rates and phase distortion is to design the amplifier with direct coupling between stages. Some of these amplifiers have offered dc coupling from the mike or phono input right to the speaker connections! To reduce or eliminate output distortion and "sticking," we need power transistors capable of quick recovery. If we are using bipolar transistors, we have got to keep them out of saturation regardless of drive. The easiest solution is, of course, to switch to power FETs, but if we are committed to using bipolar transistors, the Darlingtons and a Baker clamp makes a satisfactory solution.

Admittedly, the goals that we have outlined here for our high-fidelity audio amplifier have been a bit sparse, but in keeping with the overall theme of the book, our interest focused mainly upon the output power stages. Neither the SIT nor the short-channel MOSFETs would be particularly suitable for use in the lower-level preamplifier stages, simply because of their higher $1/f$ or flicker noise. In Fig. 7–6 we are able to compare the audio noise of a short-channel MOSFET, an equivalent-power bipolar transistor, and a small-signal J-FET. The advantages of the small-signal J-FET are obvious when driven from a high-impedance source. Before we leave this subject of low-frequency flicker noise, we might speculate that not all short-channel MOSFETs

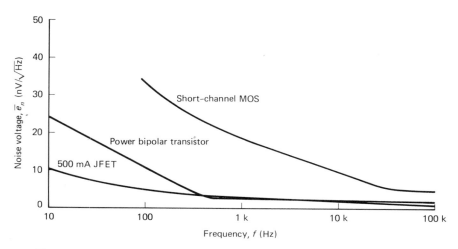

Figure 7-6 Comparison of the equivalent short-circuit input noise voltage between a typical 2-A power MOSFET, a 2-A bipolar transistor (beta = 20), and a planar epitaxy J-FET.

exhibit equivalent noise phenomena simply because of differences in material and in their fabrication. Investigators have reported on the relationship of flicker noise to orientation of the silicon upon which the MOSFET is built. For <100> we have the lowest $1/f$ or flicker noise, whereas for <111> we have the highest! This would suggest that the vertical DMOS would offer better noise performance than an equivalent-power VMOS. We might further expect that in the absence of noise-measuring equipment, we could identify noisy power MOSFETs by monitoring their drain leakage current $I_{D(off)}$.

High-frequency noise is discussed in Chap. 10.

7.4 THE SINGLE-ENDED AUDIO POWER AMPLIFIER

The simplest single-ended audio power amplifier is undoubtedly the familiar source follower, shown in Fig. 7-7 with its equivalent circuit. If we simplify the equivalent circuit by conveniently removing all parasitic capacitances except the input capacity, we can arrive at a simple expression for voltage gain:

$$\frac{e_o}{e_i} = \frac{1}{1 + \dfrac{R_1 + Z_2}{R_S(1 + Z_2 g_m)}} \tag{7.1}$$

where R_1 = total of source resistance plus any gate resistance that, for example, could result from a silicon-gate process

Z_2 = impedance of the gate-to-source and gate-to-body capacitance (there is *no* Miller effect in a source follower)

R_S = resistance in the source, which for this amplifier would more than likely be the loudspeaker

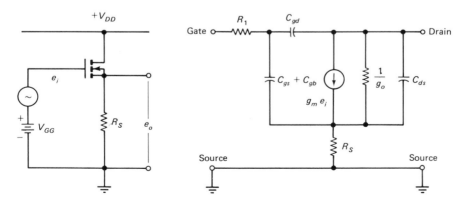

Figure 7-7 Simple source-follower and equivalent circuit of a short-channel power MOSFET amplifier.

g_m = forward transconductance of the MOSFET expressed in Mhos

We must remember that we are dealing with voltage ratios, so our calculated response must be plotted either in voltage ratio or, if we prefer, in decibels (dB) by computing $20 \log (e_o/e_i)$.

Using Eq. (7.1) we can easily calculate the frequency response of an equivalent power bipolar transistor and plot a comparison of performance as we have done in Fig. 7-8, where the comparison shows definite improvement for our power MOSFET.

We could build a common-source power amplifier wherein, although the voltage gain could be appreciable, the bandwidth would suffer as a result of Miller effect. A simple power amplifier is shown in

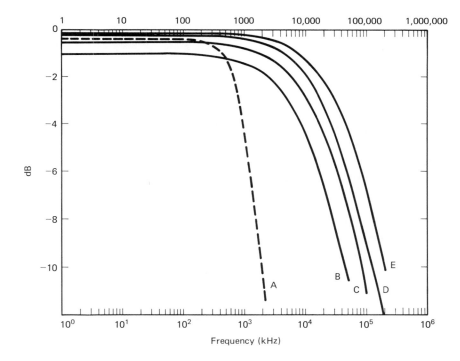

Figure 7-8 Comparison of the frequency response between power FETs and an *npn* bipolar power transistor of equivalent power. Curves: A, frequency response of an *npn* bipolar transistor of comparable size to power MOSFETs (B–E); B, response of silicon-gate power MOSFET with R_G = 60 Ω, C_{IN} = 600 pF, g_m = 1 s; C, response of silicon-gate power MOSFET with R_G = 50 Ω, C_{IN} = 800 pF, g_m = 2 s; D response of silicon-gate power MOSFET with R_G = 50 Ω, C_{IN} = 1500 pF, g_m = 5.5 s; E, response of silicon-gate power MOSFET with R_G = 50 Ω, C_{IN} = 1200 pF, g_m = 7.5 s.

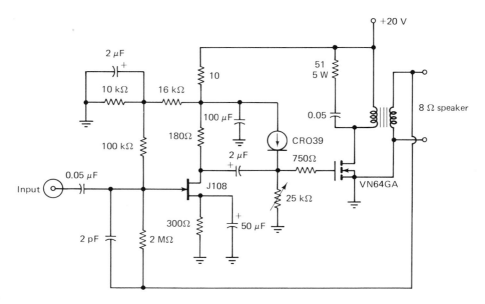

Figure 7-9 Simple power MOSFET audio amplifier. (Courtesy of Siliconix incorporated.)

Fig. 7-9, where a J-FET driver is included to complete the design. This driver circuit could also be used for our source-follower amplifier (Fig. 7-7) and feedback could be direct from the source of the MOS-FET back to the source of the driver. The bandwidth of this amplifier, even with feedback, is greatly reduced from our first example provided in Figs. 7-7 and 7-8, offering less than 2% distortion to 15 kHz and delivering between 3 and 4 W to the speaker.

Since our power MOSFETs take little drive power, we can simplify our design even further by using an op amp for the preamplifier as shown in Fig. 7-10. Gate current would be dependent upon our intended slew rate, and this, of course, would determine the drive requirements of the op amp. If we desire our amplifier to reach, for example, 50 kHz, our slew rate would be 20 V/μs. This common-source amplifier would have its input capacitance increased because of the Miller effect [Eq. (4.9)]. We can easily calculate this drive current

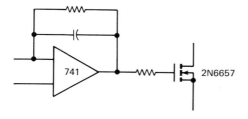

Figure 7-10 Simple power MOS-FET audio amplifier driven by an op amp.

using for our example the circuit shown in either Figs. 7-9 or 7-10. The important electrical characteristics for the 2N6657 are (taken from the data sheet)

$$\text{forward transconductance } g_m = 250 \text{ mmhos, typical}$$

$$\text{input capacitance } C_{iss} = 50 \text{ pF, typical}$$

$$\text{reverse capacitance } C_{rss} = 10 \text{ pF, typical}$$

We should be careful to note that in many instances data sheets must be carefully interpreted. For example, this JEDEC-registered 2N6657 data sheet provides the foregoing values of capacitance at zero gate-to-source voltage: in other words, with the MOSFET OFF. We obviously plan to use this power MOSFET in an amplifier with some finite bias to ensure a quiescent drain current. We should be aware that the parasitic inter-electrode capacitances of a FET move in the transition from an ON state to an OFF state, and vice versa, and it is quite important to us to know *which way they move.* Although there are a variety of cross-sectional views among the many power MOSFETs available today, we can learn from Fig. 7-11 which identifies a very marked change in many of the capacitances as the bias V_{GS} crosses the threshold V_{TH}. What is especially alarming to us is the marked *increase* in gate-to-source capacitance. However, if in our example we assume that the

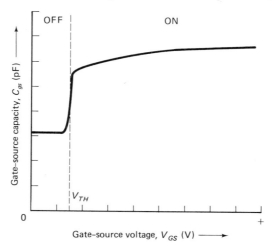

Figure 7-11 Effect of gate–source voltage upon the input capacity of an enhancement-mode MOSFET. With the channel depleted (OFF), the capacitance is at its lowest value, but with carriers flowing in the channel (ON), C_{gs} moves sharply upward. For example, a power MOS-FET with a V_{th} of typically 4.5 to 5.5 V will show an increase in C_{in} of as much as 1.8 times that value measured at $V_{gs} = 0$. This will be more fully covered in Sect. 8.4.

typical value for C_{iss} is pessimistic and the more realistic value is less, we can simply use 50 pF for a close approximation. Our gate-to-source capacitance would then be $C_{iss}-C_{rss}$, or 40 pF. Since the drain load is shown as 24 Ω, our voltage gain A_V is

$$0.250 \times 24 = -6$$

(the negative sign representing the fact that the output is 180° out of phase with the input). Using Eq. (4.9), we calculate the new value of C_{in}, which includes the Miller effect.

$$C_{in} = 40 \times 10^{-12} + [1 - (-6)] \, 10 \times 10^{-12} = 110 \text{ pF}$$

We can now calculate what our drive current will be for a 20-V/μs slew rate using Eq. (4.15):

$$\text{drive current } i = 110 \times 10^{-12}\left(\frac{20}{10^{-6}}\right) = 2.2 \text{ mA}$$

Our op amp must be able to deliver at least 2.2 mA in 1 μs if we are to

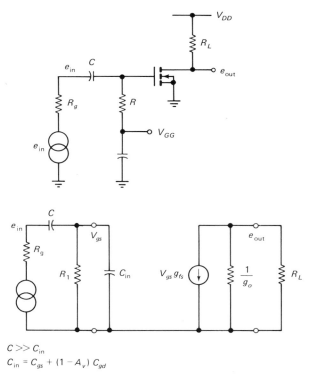

$$C \gg C_{in}$$
$$C_{in} = C_{gs} + (1 - A_v)\, C_{gd}$$

Figure 7-12 Very basic model of a common-source high-frequency audio amplifier.

provide a 50-kHz upper frequency response. Achieving 50 kHz does not come quite so simply. If we refer to our equivalent circuit in Fig. 7-7, we can further simplify it to represent a high-frequency (audio) model, as shown in Fig. 7–12, and using this model we can calculate the high-frequency cutoff point f_H, where our signal roll-off is -3 dB:

$$f_H = \frac{R_g + R_1}{2\pi R_g R_1 C_{in}} \tag{7.2}$$

From Eq. (7.2) we see that to achieve a high-frequency roll-off we need to massage the design of our circuit. Input capacity C_{in} and π are invariant, so we must work with the gate biasing R_1 and the source resistance R_g. It is not an easy task, and when options are available, the source follower is to be preferred.

7.5 THE PUSH-PULL AMPLIFIER FOR HIGH FIDELITY

These single-ended audio power amplifiers that we have just reviewed are biased Class A. Before we begin our discussion of the push-pull audio amplifier it might be well for us to review quickly the basic differences between the popular classes. The fundamental difference we find that identifies the class of operation is the *conduction angle*, that is, for what part of a full cycle (360°) drain current flows. Quite obviously, these classes are tied directly to bias. A Class A amplifier is biased to maintain a constant quiescent drain current such as to allow the full sinusoidal input signal to be reproduced at the output. For this class our conduction angle is 360°. Class B operation has the drain current flowing only during one-half of the input cycle and as a consequence only one-half of the cycle appears at the output. Our conduction angle for Class B is 180°. Class C, which is never used for audio applications, is biased well below threshold (for an enhancement-mode MOSFET) and the sinusoidal input signal is all but cut off completely at the output. This class provides a conduction angle well below 180°. To be able to visualize these classes, a representation of each is shown in Fig. 7–13.

Aside from these classes, A, B, and C, there are also intermediate classes that we can utilize in amplifier design. Possibly the most popular for push-pull audio designs is the Class AB, which simply means that the conduction angle is greater than 180° and probably less than 270°.

Amplifiers built around the vacuum tube often further subdivide the classes to include those where grid current either flowed or did not flow during part of the cycle. The suffix 2 identified grid current, whereas the suffix 1 indicated that grid current did *not* flow during any

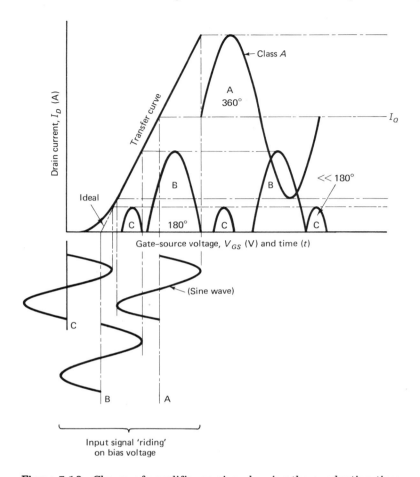

Figure 7-13 Classes of amplifier service, showing the conduction times for Classes A, B, and C. Class A is biased sufficiently high on the transfer characteristic to ensure a constant quiescent drain current and 360° conduction. Classes B and C are biased for 180° conduction (half-cycle) and less than 180° conduction, respectively.

portion of the input cycle. Using either suffix in power MOSFET audio amplifiers perhaps is meaningless, as MOSFETs are, for the most part, current-amplifier transistors that rely solely on gate voltage for drive. Yet, were we to slew an amplifier quickly as we showed in the previous example, gate current would, indeed, flow but only to charge the input capacity. For that reason, we can identify our Class AB as AB_1 without fear of contradiction.

The class of operation, whether we used Class A, B, or AB_1, determines both the distortion and the efficiency of our audio power amplifier. Theoretically at least, our Class A amplifier should offer the least distortion and the poorest efficiency. At the very best our efficiency for a Class A amplifier is limited to 50%. For small audio amplifiers such poor efficiency is not particularly bothersome, but were we interested in a high-power high-fidelity audio amplifier, we would need to make some compromises, and the first would be to boost the efficiency. This we can do by changing the class of operation.

If we jump directly to Class B operation we may also have jumped into a high-distortion problem, for a number of interesting reasons some of which we shall investigate. Our first problem with Class B is how to bias the enhancement-mode MOSFET. By definition Class B has a conduction angle of $180°$, which simply means that we cut half of the drive signal OFF. Using Fig. 7–14, where we show both an n-channel and a p-channel enhancement-mode MOSFET, we are able to observe severe distortion of the output sinusoid which is called crossover distortion. Because each transistor in Class B conducts for only one-half cycle we must ensure solid power supply rails so as not to couple audio back to earlier stages by modulating the power rails. A photograph of crossover distortion from an amplifier operating in Class

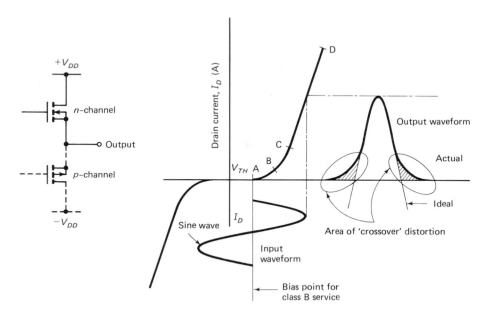

Figure 7-14   Causes for crossover distortion in Class B amplifiers resulting from the nonlinear low transfer characteristics near threshold. The area of crossover distortion shown results from subthreshold and square-law effects (see Fig. 4-7 and the accompanying discussion); it resembles that caused by storage time (see Fig. 7-5).

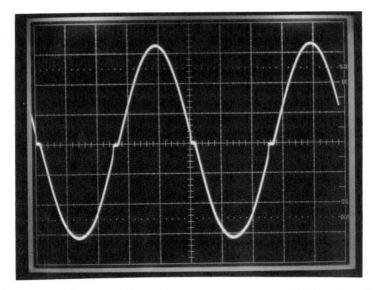

Figure 7-15 Crossover distortion on an uncompensated Class B audio amplifier.

B is shown in Fig. 7-15. To eliminate this crossover distortion we need to shift the bias to maintain operation in Class AB_1. The improvement is shown in the composite V_{DS}-I_D curve in Fig. 7-16, and the actual improvement, using the same amplifier as we used for Fig. 7-15, is shown in Fig. 7-17. For us to implement this bias shift from Class B to class AB_1 is easier said than done. Earlier we faulted the power MOSFET for its downward shift of threshold voltage with increasing temperature. To maintain our bias with the precision necessary to ensure continual operation in Class AB_1 or any other class other than Class A is no small task. We must track the temperature coefficient of the threshold voltage and adjust the bias accordingly. We cannot simply ignore crossover distortion hoping that we will not hear it; we will! Using heavy amounts of negative feedback may appear to help, but unless special precautions are taken, we only add to our problems by intensifying transient intermodulation distortion unless we have ensured both a high open-loop gain and a wide bandwidth. Or we can simply run our push-pull amplifier Class A irrespective of our desire for efficiency.

  Some have made attempts to maintain some control of threshold tracking by using diodes in the bias networks. Silicon p-n diodes, however, have a temperature coefficient closer to -2.2 mV/$^\circ$C (which, incidentally, is the same value that we find for the base–emitter junction of a bipolar transistor, and for the same reason). As a conse-

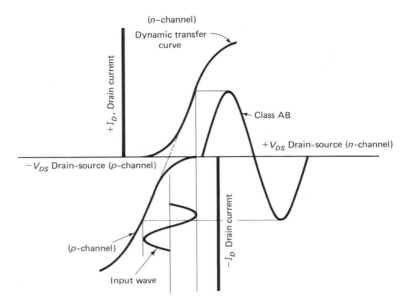

Figure 7-16 Effect of operating Class AB by the independent adjustment of bias.

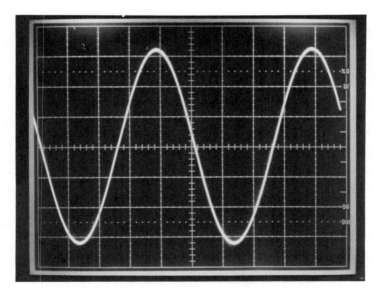

Figure 7-17 Output waveform of a properly compensated Class AB MOSFET power audio amplifier.

quence, if we select a simple silicon *p-n* diode we would be off by nearly 50%.

A better method to achieve threshold tracking might be to use a small, low-cost power MOSFET in the biasing circuitry. Since all MOSFETs exhibit a threshold coefficient between -5 and -7 mV/°C, irrespective of size or power capacity, we can achieve reasonable threshold voltage tracking over a wide temperature range using this technique. Of course, we must tie these tracking MOSFETs to the same heat sink as our power MOSFETs to ensure close temperature tracking. This method of threshold voltage tracking works equally well for any form of MOSFET amplifier, single-ended or push-pull, using either complementary pair or quasi-complementary.

Another threshold voltage tracking circuit involves a novel patented shunt diode peak clipper in a feedback network as shown in Fig. 7-18. Without the peak diode clipper the dc feedback would, in effect, undergo audio modulation that would seriously upset the drivers and subsequently result in erratic and distorted operation. The diode simply clips all the peak audio excursions and the large capacitor smooths the resulting dc, which then provides threshold bias irrespective of the temperature rise. The diode does not need to be, indeed should not be, mounted on the same heat sink with the power MOSFETs.

The illustrations that we have been reviewing decry the poor thermal property of the power MOSFET. And although there are steps that we can take to overcome this problem, there is yet quite another tack

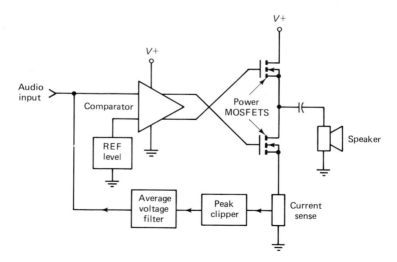

Figure 7-18 Patented feedback circuit for setting the operating class of a high-fidelity MOSFET audio amplifier. (Courtesy of Siliconix incorporated.)

open to us if we first review what is available in the marketplace. Manufacturers differ in their designs. Although many will apparently be offering carbon copies of their competitors, that is, vertical DMOS or VMOS, or perhaps planar DMOS, their fabrication may differ widely. Manufacturers differ widely both in their design philosophies and also in their understanding of customers' needs. We should be careful to exploit the advantages of particular manufacturers and not be content merely to take any power FET simply because of a cursory review of its characteristics. If we can identify a power FET whose *zero temperature coefficient*, 0 TC, lies close to the desired quiescent operating current, we should waste little time in further searching. Zero temperature coefficient is, as the name suggests, that operating point where the influence of temperature has little or no effect on performance. To illustrate the ease we might have in designing and fabricating a high-fidelity audio amplifier, we should study the design offered by Hitachi provided in Fig. 7–19. This 100-W high-fidelity audio amplifier offers an open-loop gain that peaks at approximately 15 kHz, as shown in Fig. 7–20, which would be very difficult, if not impossible for us to match were we using power bipolar transistors. The series gate resistors are to prevent spurious oscillations that might occur because of the unusually high f_t of the power MOSFETs. Ferrite beads would do equally as well. What is especially pleasant in this as well as in other power MOSFET audio-amplifier designs is the relative ease in driving the power MOSFET stages. Unlike the typical power bipolar transistor amplifier, which not only needs husky drivers but must also offer protection against "sticking" to the rails during overloads and quick shutdown to protect the finals from possible second breakdown, we can see no such need in a power MOSFET amplifier! At best we might suggest a circuit breaker in the dc line as a precaution should our loudspeaker inadvertently short the output to ground.

Many times throughout this chapter we have emphasized the importance of both a high gain and a wide open-loop bandwidth to reduce the possibility of transient intermodulation distortion (TID). A novel design using power MOSFETs is shown in Fig. 7–21. The novelty of this circuit stems from the action of the zener diodes, Z_3, Z_4, and Z_5, which prevent the *pnp* bipolar transistors, Q_{10} and Q_{11}, from conducting any time that their respective bases are driven positive by the signal. What we are witnessing that is somewhat ususual for a power amplifier is Class AB_1 operation, not in the final MOSFETs, but in the drivers. Rather than being simple differential amplifiers as we would find in most conventional designs, Q_{10} and Q_{11} have sufficient punch to handle the driving needs of the largest MOSFETs at high slew rates ($i = C\ \partial V/\partial t$). Note especially the low value of gate resistance for the power MOSFETs.

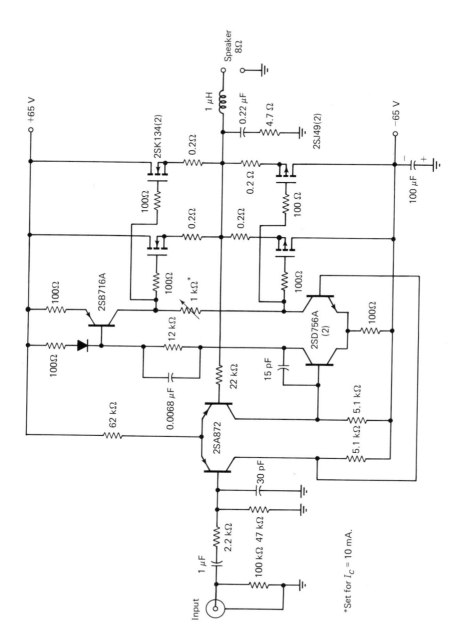

Figure 7-19 Schematic diagram of a 100-W audio amplifier using complementary pair power FETs. (Courtesy of Hitachi America, Ltd.)

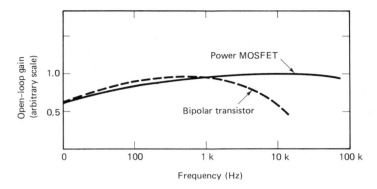

Figure 7-20   Open-loop gain of a 100-W audio amplifier. (Courtesy of Hitachi America, Ltd.)

Because of the extraordinary wide open-loop bandwidth and high gain of this amplifier, transient intermodulation distortion is *totally* absent!

7.6 THE SWITCH-MODE AUDIO AMPLIFIER

In Chap. 5 we clearly identified the remarkable switching characteristics of the power MOSFET that could result in very high efficiencies heretofore unattainable with bipolar transistors. With little modification we can translate a switch-mode converter into an efficient audio amplifier with remarkably good fidelity.

Such a high-fidelity audio amplifier is generally classified as a Class D amplifier. These amplifiers are well known for their high efficiency, but we often find that they have been largely ignored because of the difficulty in achieving high fidelity. Of course we must be careful to note that earlier Class D audio amplifiers were designed using power bipolar transistors.

We did not describe this class of operation in our earlier discussion, so it might be well for us to offer a very brief review of what Class D is and what it can do for us.

The basic principle of Class D operation may become clearer if we consider a totem-pole output stage where the active elements, the transistors, are looked upon as simple low-loss switches, as shown in Fig. 7-22. In operation we alternately close these switches at a very high rate of speed (a high frequency). The resulting duty cycle is averaged out by the series inductor in the output leg and the resulting waveform of audio appears across the loudspeaker. We can readily observe that since our transistors act simply as switches, we can have only two

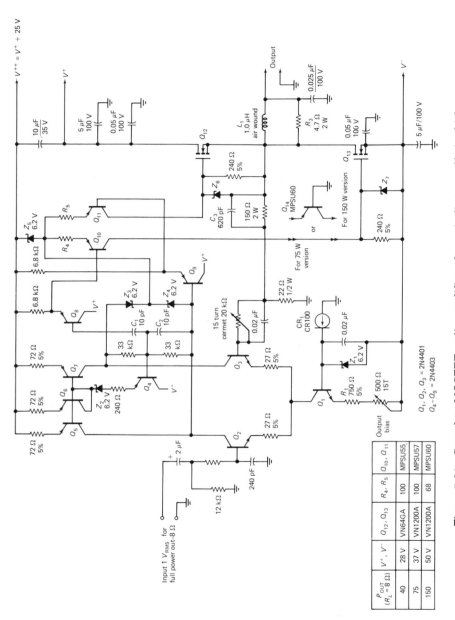

Figure 7-21 Complete MOSFET audio-amplifier schematic diagram offering high fidelity and no transient intermodulation distortion. (Courtesy *EDN*, Sept. 20, 1980 © Cahners Publishing Co. Used with permission.)

P_{OUT} ($R_L = 8\ \Omega$)	V^+, V^-	Q_{12}, Q_{13}	R_4, R_5	Q_{10}, Q_{11}
40	28 V	VN64GA	100	MPSU55
75	37 V	VN1200A	100	MPSU57
150	50 V	VN1200A	68	MPSU60

$Q_1, Q_2, Q_3 = $ 2N4401
$Q_4 - Q_9 = $ 2N4403

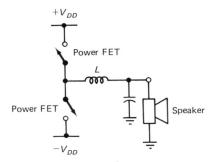

Figure 7-22 Basic operating principle of a Class D switching amplfier.

operational states, ON and OFF. We can toggle these two "switches" from a comparator acting as a PWM generator, which, in turn, we drive with a composite waveform of the desired audio modulation superimposed on a sawtooth wave at the switching frequency. The resulting PWM, as shown in Fig. 7-23, drives our totem-pole Class D switch-mode audio amplifier. We can calculate the efficiency of this switching amplifier if we know the $R_{DS(on)}$ of the power transistors and the value of our load resistance:

$$\text{efficiency} = \frac{R_{\text{spkr}}}{R_{\text{spkr}} + R_{DS(on)}} \tag{7.3}$$

If we can reduce the $R_{DS(on)}$ of our transistors to zero, we have reached 100% efficiency. However, there are problems in high-speed switching that will reduce the efficiency. We discussed these in Chap. 4 using Fig. 4-30, which also holds true most of the time for a loudspeaker (a loud-

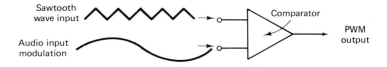

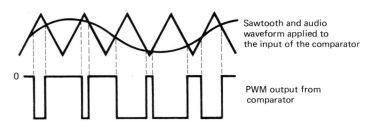

Figure 7-23   Simple procedure to develop PWM.

speaker can also appear capacitive). If we can find transistors that can switch extremely fast—and we can by using power MOSFETs that exhibit no minority-carrier storage time—and if we can select power MOSFETs with very low $R_{DS(on)}$, we can reduce both our saturation losses ($I_{rms}^2 R_{DS(on)}$) and our switching losses (as shown in Fig. 4–30) for an inductive load). However, even if our efficiencies using power MOSFETs cannot meet the efficiency of some super bipolar transistor amplifier, we can certainly outperform the bipolar transistor Class D amplifier since our power MOSFET will allow vastly faster switching speed, which, in turn, will provide much less distortion and a more pleasing sound to the critical ear of the audience.

The efficiency of a Class D audio amplifier driven by PWM is a function of both $R_{DS(on)}$ [as we saw from Eq. (7.3)] and the switching losses that incur in the power MOSFET itself (Fig. 4–30). Efficiency is *not* controlled nor is it a function of the *modulation index M*, which we can define by the equation

$$M = \left| \frac{t_{ON} - t_{OFF}}{t_{ON} + t_{OFF}} \right| \tag{7.4}$$

where t_{ON} and t_{OFF} are the ON time and the OFF time of the power MOSFETs.

Power MOSFETs offer several quite distinct advantages in the design of high-fidelity Class D switch-mode audio amplifiers. The complete absence of minority-carrier storage time is perhaps the fundamental advantage over any bipolar transistor, as it allows us to raise the switching frequency almost without limit for vastly improved fidelity, and, not incidentally, allows for easy filtering. Any switch-mode application using bipolar transistors invariably requires the use of catch diodes, but we know that with power MOSFETs we have an intrinsic parasitic drain–body–source *p-n* diode that is capable of handling the full current of the MOSFET and at a sufficient speed to offer complete protection. If we are diligent in attaining the highest efficiency, we can easily parallel power MOSFETs to further reduce their $R_{DS(on)}$.

It is important for us to use the highest possible switching frequency to reduce distortion. By its very nature PWM produces a rich field of cross-modulation products. The mixing of the switching frequency and the audio frequencies produce a plethora of what we might label even- and odd-order intermodulation products. Their mathematical relationship would appear as

$$m f_a \pm n f_m \qquad \text{and} \qquad m f_m \pm n f_a \tag{7.5}$$

where both m and n are integers, and f_a represents the audio frequency and f_m the switching frequency. We can conclude from Eq. (7.5) that it is very important to have f_m as high as possible to prevent any possibility of having these mixing products within the audio passband. If our preference is to achieve a high-fidelity audio amplifier with, say a 50-kHz upper frequency, we should endeavor to have our switching frequency certainly no lower than 500 kHz. It should come as no surprise that power MOSFETs are our only viable power transistor.

Other forms of distortion can arise if we fail to provide a suitable triangular wave in our initial generation of PWM, and, of course, we must be sure to use an op amp or comparator with a high slew rate to accommodate the switching frequency.

The output low-pass filter plays a crucial role in the performance of our Class D audio amplifier as the audio envelope—what we want to appear across the loudspeaker—is recovered by the low-pass filtering of the PWM signal. Further, to reduce the effects of the intermodulation products that we can deduce from exercising Eq. (7.5), this filter must effectively filter out all these unwanted spurious frequencies. Since we have chosen our switching frequency f_m to be at least 10 times the highest audio frequency f_a, our low-pass output filter can be designed for a cutoff frequency sufficiently high so as to provide a linear phase (flat group delay) over the audio-frequency range. Our low-pass filter must include a series inductor to recover the audio from the digital PWM. Filter design is based on a known load impedance and since a loudspeaker presents a rather erratic impedance, as shown in Fig. 7–24, we need to establish a load impedance using, for example, a series-

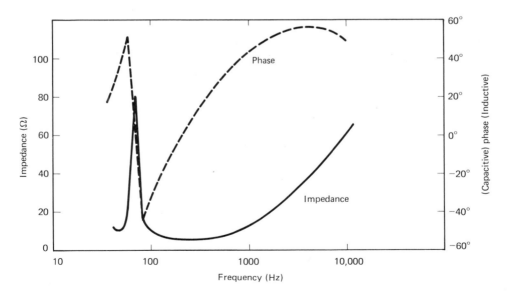

Figure 7-24 Complex impedance of a loudspeaker.

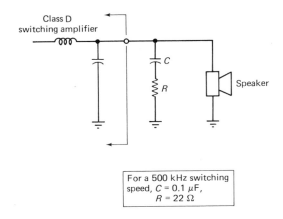

For a 500 kHz switching
speed, $C = 0.1\ \mu\mathrm{F}$,
$R = 22\ \Omega$

Figure 7-25   An RC network presents a more reasonable termination for the Class D amplifier output filter than would a loudspeaker alone.

connected RC network in shunt with the loudspeaker as shown in Fig. 7-25.

In conventional audio amplifiers we were able to improve the distortion characteristics by using moderate amounts of negative feedback. So that our transient intermodulation distortion (TID) would not be excessive, we found that this feedback should, hopefully, be limited to less than 20 dB. If we have designed the output filter to provide *flat group delay* across the audio portion, we should be able to provide negative feedback looping back to the audio input of the PWM comparator. If our filter injects phase shift with frequency, our feedback may do more harm than good.

Figure 7-26 shows a simplified block diagram of a power MOSFET Class D switching audio amplifier, and a schematic diagram is offered in Fig. 7-27.

7.7 A NOVEL IMPROVEMENT OF DYNAMIC RANGE

Contemporary music often demands dynamic range beyond what is generally available from conventional Class AB_1 audio amplifiers. Because of limitations in dynamic range, the critical listener has often had to reduce the volume and hence the output power to a fraction of the total available output power as a means of reserve for handling high-intensity sound. A team of research engineers at Hitachi Ltd. of Japan have recently announced a radical design concept whose success was guaranteed by the inclusion of power MOSFETs. This novel amplifier has been identified as operating in Class G, only to distinguish its behavior from any other amplifier.

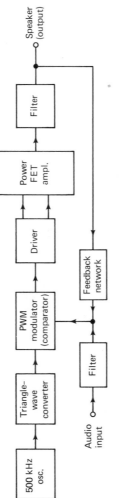

Figure 7-26 Block diagram of a Class D PWM audio amplifier.

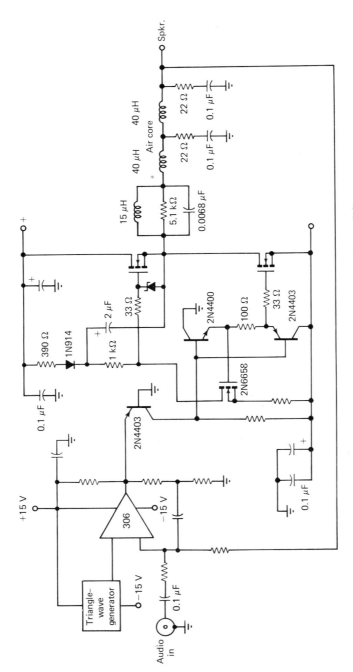

Figure 7-27 A 100-W Class D switching audio amplifier.

The fundamental problem in achieving wide dynamic range in a conventional Class AB_1 audio amplifier is simply the lack of *reserve* power. This novel Class G power amplifier operates much the same as a conventional Class AB power amplifier using bipolar power transistors. However, when we need an extra boost of power, the audio switches in a reserve using a combination of logic and a totem-pole of both power bipolar transistors and power MOSFETs. We perhaps can achieve some understanding of how this unusual amplifier operates by following a *brief* explanation of Fig. 7–28, where we have one-half of a totem-pole output stage consisting of a Darlington pair and a power MOSFET. The voltage source, V3, prevents the Darlington from entering into saturation and thus preserves the high-speed performance necessary for high-fidelity reproduction. D_2 protects Q_4 from excessive reverse voltage. Since it is important that Q_2 react quickly to the incoming signal, a small value of resistance is bridged across from gate to source, and this forces us to add an emitter follower, Q_4, to speed up Q_2 by providing additional current to charge C_{in}. The incoming audio signal voltage drives the Darlington pair much the same as the conventional Class AB_1 amplifier, and collector voltage is derived from V_{c1} through D_1. Transistors Q_4 and Q_2 are OFF by virtue of the fact that V_{c1} has biased the emitter of Q_4 higher than its base bias. This situation

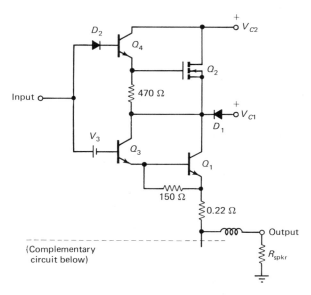

Figure 7-28 One-half totem pole of an elementary Class G audio amplifier. (© 1978 IEEE. Reprinted, with permission, from "Highest Efficiency and Super Quality Audio Amplifier Using MOS POWER FETs in Class G Operation," by Tohru Sampei et al., which appeared in *IEEE Trans. Consumer Electronics*, Vol. CE-24, No. 3, Aug. 1978.)

remains until we drive a sufficiently large audio signal voltage into the base to raise the base sufficiently to begin conduction. At that moment we have Q_4 turning Q_2 ON and D_1 becomes reverse-biased by V_{c2}. Our Class G amplifier now has the reserve power contributed by the higher supply voltage, V_{c2}.

7.8 SUMMATION OF BENEFITS AND PROBLEMS IN USING POWER FETS

Aside from that unusual biasing problem with the SIT, where a positive gate-to-source bias causes a reversion to bipolar characteristics, power FETs and in particular power MOSFETs apparently suffer from only one problem, the need to operate at a high quiescent drain current. Earlier in their development we were also concerned with their high $R_{DS(on)}$, but improved design has now lowered their effective V_{SAT} to equal the best bipolar transistors.

Their advantages, however, far outweigh these few disadvantages. No audio amplifier is easier to drive and because of no minority-carrier storage time, "sticking" is literally impossible since they cannot saturate. Their nearly linear transfer characteristics aid materially in reducing higher-order distortion products. During hard driving where a bipolar transistor's beta would fall off with increasing collector current, the power FET's transconductance stays firm. Finally, we are able to parallel them ad infinitum.

REFERENCES

COBBOLD, RICHARD S. C., *Theory and Applications of Field-Effect Transistors.* New York: Wiley-Interscience, 1971.

DUTRA, JOHN A., "Digital Amplifiers for Analog Power," *IEEE Trans. Consumer Electronics*, CE-24, No. 3 (Aug. 1978), 308–17.

GREENBURG, RALPH, ed., *Motorola Power Transistor Handbook.* Phoenix, Ariz.: Motorola Semiconductor Products Division Inc., 1960.

MANDL, MATTHEW, *Fundamentals of Electronics*, 3rd ed. Englewood Cliffs, N.J.: Prentice-Hall, Inc., 1973.

SAMPEI, TOHRU, and SHIN-ICHI OHASHI, "100 Watt Super Audio Amplifier Using New MOS Devices," *IEEE Trans. Consumer Electronics*, CE-23, No.3 (Aug. 1977), 409–17.

SAMPEI, TOHRU, et al., "Highest Efficiency and Super Quality Audio Amplifier Using MOS Power FETs in Class G Operation," *IEEE Trans. Consumer Electronics*, CE-24, No. 3 (Aug. 1978) 300–307.

STONE, ROBERT T., and HOWARD M. BERLIN, *Design of VMOS Circuits with Experiments.* Indianapolis, Ind.: Howard W. Sams & Company, Inc., 1980.

eight

Using the Power FET in Logic Control

8.1 INTRODUCTION

Perhaps coincidentally, maybe even providentially, the introduction of the power FET followed closely that of the microprocessor. Today they both are affecting almost every area of our life, and for good reason. Never have we enjoyed a simpler or easier means to attach brawn to brain than by the marriage of these two technologies.

The n-channel, enhancement-mode power MOSFET, with its positive threshold voltage and its ultrahigh input resistance, makes possible the direct interface with nearly any driver capable of a positive output voltage. Whatever we use for a driver, whether it be DTL, RTL, TTL, or even CMOS, if it is capable of outputting a positive voltage swinging from a voltage less than threshold to, say, 5 to 30 V, we can with this logic control nearly unlimited power. Because of the very high input (gate) resistance of the power MOSFET, we can parallel, almost without limit, many power MOSFETs until we achieve the current handling necessary to perform any job. We are, of course, limited in this exercise by the switching speed for, as we have read in earlier chapters, our gate current is dependent upon both the magnitude of input capacity and the switching speed [see Eq. (4.15)]. Consequently, the faster we wish to switch the MOSFET, the greater the gate current needed to drive the power FET.

Power MOSFETs controlled directly from logic can be found in industrial test equipment, telephone line switching, a variety of computer and peripheral equipments, more and more in household appliances, and as reliability improves, we shall find power MOSFETs in life-support equipment. Microprocessor-controlled power MOSFETs are already in American-made automobiles.

As computers proliferate so will power MOSFETs. They go together. Now we can take the near limitless capacity for artificial intelligence, couple it to our power MOSFET, and with that combination we can literally let them work for us.

In this chapter we examine the characteristics of a power MOS-FET as a logic-driven switch; we get an appreciation of how the bootstrap makes driving our MOSFET switch more practical for grounded loads. We consider the problems and cures of driving a variety of loads, especially highly reactive ones that might destroy a bipolar power transistor. High-speed line drivers are discussed in detail and of course, of utmost importance we learn how to interface power MOSFETs to all kinds of logic.

8.2 GROUND RULES FOR DRIVING THE POWER MOSFET

One thing for certain: we are going to discover that it is easy to interface power MOSFETs to logic but that there are some very fundamental rules. If we are obedient to these rules—and there are not many of them—we will get along very well. Interestingly, we have been exposed to most of these rules in earlier chapters, so for the most part we will not feel hemmed in by reviewing them now.

First, we know, if only intuitively, that if we exceed the ratings of our power MOSFETs we are likely to damage them. In Chap. 4 we discussed in some detail the *safe operating area* of power transistors. Here we learned that the ratings on most data sheets must be carefully weighed. We do not simply impress the maximum operating voltage across a power MOSFET and expect it to control the maximum current. Not simultaneously, in any case. Neither should we exceed either the maximum drain-to-source voltage or the maximum drain current. This should raise a flag for those of us who are thinking of using power MOSFETs to drive inductive loads. Driving an inductive load has quite the opposite effect from driving a capacitive load. Where with the capacitive load we have a high inrush of current, when we remove the drive from an inductive load we have a large voltage buildup. The former inrush of current stems from the capacitor acting at zero charge as a short circuit; the latter high voltage buildup results from the collapsing magnetic field inducing a counter-EMF in the windings of the inductor. What we have to be on guard is that this counter-EMF does not build to a level higher than the breakdown voltage of our power MOSFET.   Although this counter-EMF builds but for a very brief time and although our power MOSFETs exhibit excellent energy capability, nonetheless it is just not good design to expect the power MOS-FET to withstand this shock of energy at each switching action. Inductive switching also holds another problem which we also discussed in Chap. 4. The waveform of the counter-EMF may show voltage leading current, which can result in switching losses within the MOSFET (see Fig. 4-30).

Our cure for inductive switching overvoltage is to shunt the inductive load with a fast diode that will absorb the transient voltage spike. We must be careful to ensure that the diode will respond swiftly, so if we select a gold-doped diode we should be safe from transients. Frequently, a series resistor and diode combination are used to shunt the inductor as shown in Fig. 8-1. There are, of course, other ways we can

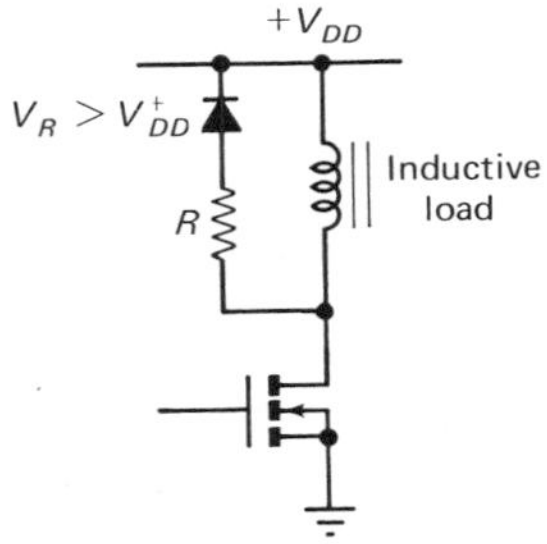

$$+V_{DD}$$

$$V_R > V_{DD}^+$$

R

Inductive load

Figure 8-1 Simple "free-wheeling" diode to catch the counter-EMF resulting from the collapsing field when the power FET turns OFF. (Courtesy of Siliconix incorporated.)

protect the transistor from excessive overvoltage. Let us assume that we are using a power MOSFET as a solenoid driver in some type of appliance, perhaps a microprocessor-controlled dish washer. Our power mains are, for the sake of economy, raw dc taken from a simple rectifier in the ac mains. In almost every situation, transients exist. Numerous studies by some public utilities have shown that such transients can, perhaps infrequently, soar to twice the ac line voltage. If our appliance is not protected with a TransZorb®[1] we then should be careful to protect our power MOSFETs. A very acceptable overvoltage protector that doubles as a protector from the counter-EMF from the inductor is a zener diode placed across the power MOSFET as shown in Fig. 8-2.

There are some ground rules for us to observe when we use power MOSFETs to drive highly capacitive loads. A capacitor without an electrical charge appears as a short circuit, so immediately we can easily

[1] TransZorb is a registered trademark of General Semiconductor Industries, Inc.

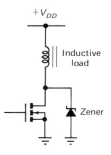

$+V_{DD}$

Inductive load

Zener

Figure 8-2 Zener diode affords picosecond response time for sure-fire protection. (Courtesy of Siliconix incorporated.)

recognize that to drive a capacitive load we are going to have to dissipate some power. Work needs to be done and we can easily calculate how much by using this well-known equation:

$$W = \tfrac{1}{2}CE^2 \qquad \text{watt-seconds} \qquad (8.1)$$

We want to be sure that this power is not dissipated within the power MOSFET. We learned in Chap. 4 that power FETs as a general rule do not suffer from the phenomenon called second breakdown. Yet we must be alert to a failure mode that looks so much like second breakdown that we could argue quite convincingly that power FETs are susceptible to this failure mechanism. This failure mode results from a form of thermal runaway. Although we may have a thermally degenerate power MOSFET, we can, nevertheless, witness thermal runaway by sinking a high current through our power MOSFET that raises the chip temperature. Remember that $W = I_D^2 R_{DS(on)}$. This problem is not confined solely to capacitive loads; any load susceptible to high inrush currents requires our close attention.

The problem becomes acute when two situations develop. First, our load demands high inrush current; and second, we are driving our power MOSFET from low voltage logic, say for example, TTL. What we are doing is limiting the turn-ON of our power MOSFET. In other words, we are limiting our current flow with excessive $R_{DS(on)}$, and that is potentially dangerous. We could exceed our SOA for the power MOSFET.

What we should try to do if we are faced with high inrush current is to increase our gate voltage to ensure that our power MOSFET is operating in its ohmic region, and, of course, we should avoid slow ramping up of our gate driving voltage. Perhaps one of the most severe loads requiring horrendous inrush currents is the common incandescent light bulb. A 100-W incandescent measured 10 Ω and a 250-W flood measured only 2½ Ω when cold. Although these lamps would heat up rapidly and their resistances would rise, we would, nevertheless, witness considerable inrush current if only momentarily.

Our final ground rule that we must consider before we begin our study is perhaps of more immediate concern because we are addressing how to use a power MOSFET when our driving voltage *cannot* tie directly between gate and source. Previously, we assumed that if our driver delivered 10 V, then our gate-to-source voltage was 10 V and our power MOSFET reacted accordingly by turning ON.   An efficient switch, as we wish our power MOSFET to be, must offer low resistance, for our power MOSFET $R_{DS(on)}$ is proportional to the gate-to-source voltage.   If our driver delivers 10 V but the gate-to-source voltage is only, say, 3 V, then the $R_{DS(on)}$ rises and we discover much to our chagrin that our power MOSFET switch is quite inefficient.

In today's market many of the available power MOSFETs operate with a logic-compatible threshold voltage ranging between 1 and 2 V. The greater majority offer maximum drain current and minimum $R_{DS(on)}$ with a gate-to-source voltage around 15 V.   Therefore, if we are to exploit the advantages of these power MOSFETs, we must manage to drive the gate with a voltage that can rise to, say, 15 V *above* the source voltage.   We can understand our dilemma perhaps somewhat more vividly if we consider the circuit in Fig. 8-3.   A positive drive voltage begins to turn the power MOSFET ON, and as current flows a voltage builds across the source resistor R_S, which immediately reduces the gate-to-source voltage, which, in turn, reduces the drain current, which, in turn, reduces the voltage across R_S.   In other words, we cannot turn the power MOSFET ON sufficently to do any worthwhile work.

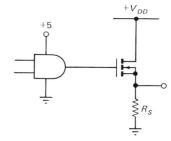

Figure 8-3 Paradox of a logic-driven source follower.   As the logic output ramps upward, the source voltage follows, resulting in a very degenerate condition.

8.3 DRIVING THE POWER MOSFET

We drive our power MOSFET so that it might perform the task we have in mind.   We can drive it to turn ON fast or slow; to pass high current or low current.   The drive power that we will need will depend on whether we want to switch fast or slow and on how much or how little current we wish to pass.   Although the dc resistance of a power MOS-FET's gate is many megohms, we must remember that we have a shunting capacity C_{in} that plays a major part in the establishment of our driving circuits. At high speeds this input capacity becomes a highly reactive load that demands our careful attention.

Before we involve ourselves further into the problems and solutions of driving the power MOSFET we had better take a fresh look at the power MOSFET we may be considering for our job. Several manufacturers of small power MOSFETs have added gate protection zeners that limit the gate voltage to protect the oxide from inadvertent punch-through that might occur either through static discharge or from the application of excessive gate voltage. In Chap. 4 we thoroughly discussed the fundamental problem associated with this so-called zener, which we do not need to repeat here other than to remind the reader *never* to allow the gate-to-source voltage to swing negative, even for a moment. If we think that our driver will output any sort of negative overshoot, we should select a power MOSFET without zener protection. Another precaution that does not need further elaboration is the high figure of merit typical of the power MOSFET. If we compare any power MOSFET with a comparably rated bipolar transistor, we will see immediately that the power MOSFET is a high-frequency transistor. Many of the fundamental advantages of our power MOSFET arise from features such as the total absence of minority carriers. Although we may not be used to working with high-frequency transistors in logic control applications, we do have to follow a few basic rules. Our first rule is to keep the leads to the power MOSFET short. This rule applies not only to the gate but also to the source. If we plan to use common-source, then by all means we should endeavor to ground the source as soon as possible. We cannot overemphasize short leads. A ¼-in. length of hookup wire can resonate at surprisingly low frequencies when properly loaded by a capacitor which could easily be our MOSFET's input capacity. We also need to reduce the feedback that might arise from excessive source lead inductance. In some situations where we are pushing the performance of our power MOSFET to higher and higher speeds, quite often we find definite improvement by selecting a power MOSFET designed especially for high-frequency applications simply because these high-frequency devices have multiple source leads to reduce package inductance. If we are stuck having to use long gate leads, we can reduce the potential for trouble by either stringing a ferrite bead near the gate or by placing a small value of resistance in series with the gate. A value of 50 to 100 Ω is generally adequate but, of course, the higher the better if it does not affect the circuit. In some circumstances we may choose a power MOSFET with a silicon gate, and this may provide sufficient series resistance so as not to require an external resistor.

Occasionally, when driving the power MOSFET from a high-impedance source there is a possibility of inducing positive feedback that might force our circuit into oscillation. Our second rule follows the first rule; that is, we should "dampen" the gate impedance by adding either a small ferrite bead close to the gate terminal or a small value of series resistance in the gate lead.

The static input impedance of our power MOSFET is nearly infinite, whereas the dynamic input impedance varies according to either the frequency or switching speed. In many of our applications involving logic control, we are extremely interested in speed, which forces us to examine the charge characteristics of the power MOSFET's input.

8.4 A TRANSIENT ANALYSIS OF A POWER MOSFET[2]

When we drive the power MOSFET as a high-speed switch we must remember that the dynamic characteristics of the gate depend on how fast we wish to turn the MOSFET ON or OFF. Our situation differs only in magnitude from the problems associated with driving a capacitive load with a power MOSFET. If the MOSFET's equivalent input capacity C_{in} is known and the required change in gate voltage is known (from a knowledge of the transfer characteristics and the desired drain current), we are able to compute the energy needed by using Eq. (8.1), which we can now rewrite as

$$W = \frac{1}{2} C_{in} \Delta V_{gs}^2 \qquad (8.2)$$

We will be referring to various parameters using lowercase for subscripts, such as V_{gs} in Eq. (8.2) to represent *dynamic* values, and when applicable, we will revert to uppercase letters for static values, such as drain current, I_D. As we saw in Chap. 4, C_{in} is dependent upon both V_{DS} and V_{GS} and to some extent upon V_T. As we move our power MOSFET from the OFF state to the ON state we see a dramatic change in capacity. Since input capacity is, at best, a difficult variable to lay hold of, a better method is to determine the gate charge Q_g as a function of V_{gs}. Our equation then becomes

$$W = \frac{1}{2}(\Delta Q_g)\Delta V_{gs} \qquad \text{watt-seconds} \qquad (8.3)$$

As we follow our discussion of the dynamic input characteristics, we will reach a better understanding if we use an actual power MOSFET for our example. In Fig. 8-4 we have the charge characteristics of a Siliconix power MOSFET, type VN64GA. Accompanying the charge

---

[2]This section has borrowed heavily from A. Evans and D. Hoffman, "Dynamic Input Characteristics of a VMOS Power Switch," *Siliconix Application Note AN79-3*, Copyright 1980. Used with permission.

Electrical characteristics (25° C unless otherwise noted) Siliconix VN64GA								
		Characteristic	Min.	Typ.	Max.	Unit	Test Conditions	
1	S T A T I C	BV_{DSS}	Drain-source breakdown	60			V	$V_{GS} = 0, I_D = 500\ \mu A$
2		$V_{GS(th)}$	Gate threshold voltage	1.0	2.7	4.0		$V_{DS} = V_{GS}, I_D = 10\ mA$
3		I_{GSS}	Gate-body leakage		0.3	100	nA	$V_{GS} = 12\ V, V_{DS} = 0$
4		I_{DSS}	Zero gate voltage Drain current			500	μA	$V_{DS} = $ max. rating, $V_{GS} = 0$
5		$I_{D(on)}$	ON-state drain current	12.5			A	$V_{DS} = 25\ V, V_{GS} = 12\ V$
6		$R_{DS(on)}$	Drain-source ON resistance		0.3	0.4	Ω	$V_{GS} = 12\ V, I_D = 10\ A$
7	D Y N A M I C	g_{fs}	Forward transconductance	1.5	2.2		$\mho$	$V_{DS} = 20\ V, I_D = 5\ A$
8		C_{iss}	Input capacitance		700			$V_{GS} = 0, V_{DS} = 25\ V,$ $f = 1.0\ MHz$
9		C_{rss}	Reverse transfer capacitance		25		pF	
10		C_{oss}	Output capacitance		325			

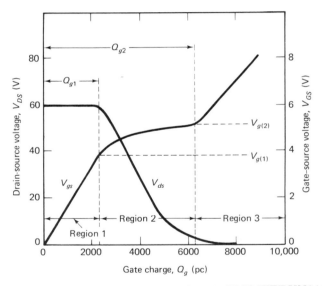

Figure 8-4 Charge characteristics of the Siliconix VMOSFET VN64GA. (Courtesy of Siliconix incorporated.)

characteristics is a tabulation of the electrical characteristics of the VN64GA taken directly from a data sheet. From our observations of Fig. 8-4, we can detect three distinct states or regions. In region 1, V_{gs} is below the threshold voltage and our MOSFET is OFF. In region 2, V_{gs} rises above the threshold voltage and drain current I_D commences to flow. This region is identified as $V_{g(1)}$. At region 3, V_{ds} saturation is reached and this region we label $V_{g(2)}$. Any further increase in the gate-to-source bias makes no further improvement in either $R_{DS(on)}$ or $V_{DS(SAT)}$.

We can now calculate the effective input capacity C_{in} for each region by using the information contained within the graph shown in Fig. 8–4, using the equation

$$C = \frac{Q}{E} \tag{8.4}$$

In region 1, C_{in} is constant by virtue of the constant slope of V_{gs} versus Q_g. The approximate value of C_{in} may be calculated as

$$C_{in(1)} = \frac{Q_{g(1)}}{V_{g(1)}} = \frac{2450 \text{ pC}}{3.8 \text{ V}} = 645 \text{ pF}$$

In region 2 things begin to happen, for as the bias rises beyond threshold, drain current begins to flow, the transconductance increases and with the increasing transconductance we have an increasing gain, and the Miller effect causes C_{in} to increase as we saw earlier in Eq. (4.9). We can calculate the approximate value of C_{in} for region 2 taking the values from our graph

$$C_{in(2)} = \frac{6250 - 2450}{5.1 - 3.8} = 2923 \text{ pF}$$

When we reach region 3, our MOSFET is full ON, $R_{DS(on)}$ has reached its asymptote where further reduction is impossible, $V_{DS(SAT)}$ is also at its minimum, and the Miller effect ceases to affect C_{in}. Because of the low $V_{DS(SAT)}$, the depletion fields have constricted and we calculate from our graph a new and somewhat higher value of C_{in} than we witnessed for region 1.

$$C_{in(3)} = \frac{8000 - 6250}{7.1 - 5.1} = 875 \text{ pF}$$

$C_{in(1)}$ and $C_{in(3)}$ correspond to C_{iss} and are approximately equal to $C_{gs} + C_{gd}$. It is understandable that they differ in magnitude since we see that $V_{DS(SAT)}$ and V_{DG} also differ between region 1, where the MOSFET was OFF, and region 3, where the MOSFET is full ON. Between these two states we see that C_{gd} has undergone a large change as shown in Fig. 8–5, where the bias conditions between regions 1 and 3 are compared. In region 1 the drain area under the gate is depleted of carriers (increased depletion area), which lowers the value of C_{gd}, and in region 3 this area becomes flooded with carriers, which raises C_{gd}. We can see that our calculations for $C_{in(1)}$ are close by comparing our calculated value (645 pF) with the typical value of C_{iss} taken from the data sheet for the VN64GA (see Fig. 8–4).

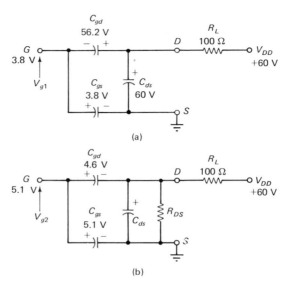

Figure 8-5 Bias conditions for C_{gd}, C_{gs}, and C_{ds} for the Siliconix VN64GA. (a) Bias condition at end of region 1. (b) Bias conditions at start of region 3. (Courtesy of Siliconix incorporated.)

In reexamining Fig. 8-4, we see in region 1 that a finite charge accumulation is necessary before V_{gs} reaches the threshold voltage V_T. To charge C_{in} takes a finite amount of time which we can identify as *turn-ON delay time*. Once we reach the threshold voltage, any further increase in V_{gs} contributes to turning ON the power MOSFET. This we identify as *turn-ON time*. Turn-ON time ceases once $V_{DS(SAT)}$ occurs at $V_{g(2)}$.

Although we know that FETs are free of minority-carrier storage time, we can experience what we might call a *turn-OFF delay time*, not caused by minority-carrier saturation but, instead, caused by excessive charge on $C_{in(3)}$. In other words, an overdrive situation will cause turn-OFF delay! This can be avoided by careful design and equally careful biasing, but probably the benefits are not worth the effort. We can use the values of charge to derive a first-order approximation of our anticipated turn-ON delay time as well as the turn-ON time if we first assume that our gate is being driven by a constant-current source, I_g, using the equation

$$t = \frac{Q_g}{I_g} \tag{8.5}$$

The values of $Q_{g(1)}$ (turn-ON delay time) and $Q_{g(2)}$ (turn-ON time) can be taken directly from Fig. 8-4. However, we generally use logic drivers that appear as resistive sources, which negates Eq. (8.5) and forces us to consider the effects of an RC time constant. Such equations have been derived where $C_{in(1)}$ and $C_{in(2)}$ (the latter being the input capacity contributed by the Miller effect) are constants.

$$t_1 = \frac{Q_{g(1)}}{V_{g(1)}} R_{gen} \ln \left[1 - \frac{V_{g(1)}}{V_{GG}} \right] \qquad (8.6)$$

and

$$t_2 - t_1 = - \frac{Q_{g(2)} - Q_{g(1)}}{V_{g(2)} - V_{g(1)}} R_{gen} \ln \left[1 - \frac{V_{g(2)} - V_{g(1)}}{V_{GG} - V_{g(1)}} \right] \qquad (8.7)$$

where V_{GG} is the driver open-circuit voltage and R_{gen} is its output resistance.

Continuing our example using the Siliconix VN64GA, let us assume that our driver can output +10 V with an output resistance R_{gen} of 10 K Ω. From Fig. 8-4 we get

$$Q_{g(1)} = 2450 \text{ pC} \qquad V_{g(1)} = 3.8 \text{ V}$$
$$Q_{g(2)} = 6250 \text{ pC} \qquad V_{g(2)} = 5.1 \text{ V}$$

Substituting in Eqs. (8.6) and (8.7), we get

$$t_1 = 3.08 \text{ } \mu s \qquad t_2 - t_1 = 6.88 \text{ } \mu s$$

Our turn-ON time is simply t_2 or approximately 10 μs. As we reduce R_{gen} our times are reduced in direct proportion.

We can calculate turn-OFF times in a similar manner; however, we begin with the MOSFET fully ON and V_{gs} at V_{GG}. Our *turn-OFF delay* lasts until V_{gs} drops to $V_{g(2)}$, at which time V_{DS} rises out of saturation, the Miller effect recedes, and once V_{gs} drops below $V_{g(1)}$, the power MOSFET is completely OFF.

Our discussion to this point has focused upon charge transfer but not on drive current, which we recognize is necessary to switch these power MOSFETs quickly. It is easy for us to calculate the required drive current by using Eq. (8.5) and substituting known values for charge and time. Using our example, we need 6250 pC to charge the VN64GA gate to $V_{g(2)}$, which, in turn, will drive the output to saturation.

$$I_g = \frac{6250 \times 10^{-12} \text{ C}}{20 \times 10^{-9} \text{ s}} = 313 \text{ mA}$$

In subsequent sections in this chapter we show a variety of methods by which we will be able to drive power MOSFETs of all sizes and capacities.

What we have just discussed assumed that our power MOSFET was switching a pure resistive load. In many applications that we shall face we are going to have relays, solenoids, and various types of transformers, all of which present a rather healthy inductive reactance to the drain of our power MOSFET. Earlier in this chapter we saw the wisdom of clamping our inductive loads to prevent any accident that might result from overvoltage, so in the discussion to follow we assume that a clamp diode shunts the inductive load. A simple circuit used for illustration for our discussion is shown in Fig. 8-6. In this schematic we show a speed-up RC network which we shall examine later in this

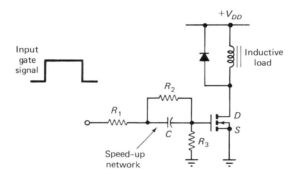

Figure 8-6 Typical switching circuit with a clamped inductive load.

chapter. Earlier in Chap. 4 we saw that an inductive load causes a delay in ON current and a delay of OFF voltage during the switching action. This is clearly seen in Fig. 8-7. These waveforms are quite idealistic, as we have taken several liberties; for example, we have assumed that current flow through the inductive load is continuous—through the MOSFET when ON and through the clamp diode when OFF. We have further assumed that our circuit has no stray inductances, so that whenever the power MOSFET turns ON it simply commutes the clamp diode current. Using the various notations offered in Fig. 8-7 for switching times, such as $t_{r(I_D)}$ to mean *rise time for drain current* and $t_{f(V_D)}$ for *fall time for drain voltage*, we can compare the typical relationships for an International Rectifier high-voltage power MOSFET, type IRF330[3], in Fig. 8-8.

---

[3] The IRF330 is also available as a 2N6760.

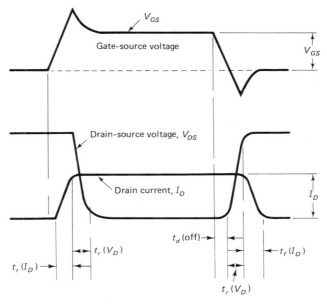

Figure 8-7 Idealized switching waveforms for clamped inductive load. (Courtesy of International Rectifier, Inc.)

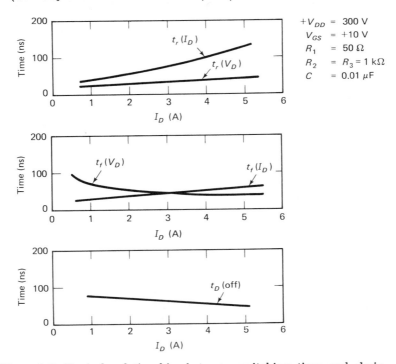

Figure 8-8 Typical relationship between switching time and drain current with a clamped inductive load and a gate speed-up circuit feeding an International Rectifier power MOSFET, IRF330. (Courtesy of International Rectifier, Inc.)

There are three different modes of operation for power MOSFETs that are described in this and subsequent sections. The first mode that we will address is the common-source configuration, which is, by far, the easiest to drive. The other modes that we cover include the common-drain, or source follower, and the totem pole; the latter we covered to some extent in previous chapters. In our introduction to this chapter we, perhaps somewhat glibly, declared that power MOS-FETs can be driven directly from DTL, RTL, TTL, or CMOS. We should now admit that we need to further qualify this statement. True, n-channel enhancement-mode power MOSFETs can be driven from these logic elements *provided* that (1) our power MOSFET has a threshold voltage sufficiently low so as to be compatible with both the logic element's *high* state and its *low* state; and (2) our logic element can handle the current required to switch the power MOSFET in the required time. Let us pause to expand on these two qualifiers. First, we must choose our power MOSFET so that when our logic element commands that the FET turn ON, the power MOSFET indeed does. When the logic element issues the command to turn OFF, again the MOSFET follows. This means that if we decide to use DTL, RTL, or TTL, we must be fully OFF at a gate voltage of 0.8 V or less, and ON with a gate voltage above approximately 2.4 V. However, the question remains: How far ON? Here we have options available that allow us some adjustment as to *how far ON*. We cover this in great detail later in this chapter. The second qualification involves the speed at which we want our power MOSFET to switch. We saw in Sec. 8.4 that to turn ON a Siliconix VN64GA, which typically has 645 pF of input capacity, takes about 300 mA of gate current for a 20-ns turn ON. We have to be sure that our logic element can supply that much drive current in that short a time. Again we have options, which we also cover in this and subsequent sections in this chapter.

8.6.1 Using Logic to Drive the Common-Source Power MOSFET

By now we are well aware how to turn ON a power MOSFET. It takes a positive gate voltage V_{GS} between gate and source to activate an n-channel MOSFET; and a negative gate voltage $-V_{GS}$ between gate and source to switch a p-channel MOSFET. The latter does not find much application as a stand-alone common-source logic-driven element since logic output is generally positive. Since we are addressing direct logic drive, we will concentrate our attention on the n-channel MOS-FET. What makes the common-source power MOSFET so easy to drive

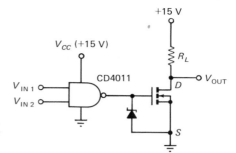

Figure 8-9 CMOS gate driving a logic-compatible power MOSFET with zener gate protection.

from logic can best be understood by refering to Fig. 8–9. The full output voltage of the logic element appears as V_{GS}. If we placed a resistor in the source lead of the power MOSFET, we would defeat the advantage of common-source operation, as the full output voltage of our logic element would not appear as V_{GS} but rather as $V_{GS} - I_D R_S$, where I_D is variable dependent upon the *effective* V_{GS}. We are jumping ahead of our present topic; we discuss the common-drain or source-follower circuit in detail later and offer solutions to this seemingly perplexing problem.

What is generally well known as high-voltage CMOS is ideal for driving power MOSFETs because we can operate with supplies of ±15 V, and there are few power MOSFETs that will not be turned ON hard with +15-V gate drive.   The simple circuit in Fig. 8-9 is typical of a CMOS drive direct-coupled to a power MOSFET; nothing could be quite so simple.   We do have one problem if we use a CMOS driver and that becomes a problem only if we desire to switch fast.   A CMOS gate, for example, a 74C00, is limited in the amount of current it can handle. If we are determined to use CMOS drivers and we desire a higher-speed switch, we do have a few options.   If one CMOS gate slows us down because of its limited current-handling capacity, two CMOS gates tied in parallel will double the current and four CMOS gates will quadruple the current.   So our first option to increase the switching speed of our power MOSFET is to drive the gate with as many paralleled CMOS gates as is practical.   Since quad CMOS gates are common, we might find this option the most satisfactory.   Because of the characteristics of the CMOS gate, we should be careful to see that we operate it as close to the ±15 V rails as possible to improve its current handling, and this is as important for one CMOS gate as it is for a quad.

Referring back to Fig. 8–9, we can label the components and then see how the switching performance compares as we manipulate first the rail voltage for the CMOS driver (Fig. 8–10) from 15 to 10 V.   Note especially that as the drive current drops (V_{CC} = 10 V) the Miller effect swamps the power MOSFET's input capacitance and forces a slow

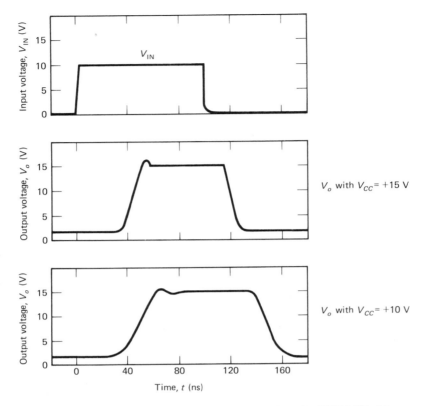

Figure 8-10 Switching performance of the Siliconix VN66AF driven by a CD4011 CMOS gate (see Fig. 8-9). (Courtesy of Siliconix incorporated.)

rise time.   An improved response time, especially in rise time, is evident in Fig. 8–11, where we have a quad CMOS gate driving the power MOSFETs. In these illustrations we used the Siliconix VN66AF power MOSFET, a 60-V, 2-A n-channel MOSFET with a nominal input capacity of 50 pF.

If extra CMOS gates are unavailable, we can then use a very popular emitter-follower circuit that is found in many applications. This circuit appeared in Chap. 5 (see Fig. 5–10), where we used it to speed the drive to a high-input-capacity power MOSFET. The basic emitter-follower MOSFET driver is shown in Fig. 8–12. Using this circuit the current delivered to the power MOSFET is the output current of the CMOS gate multiplied by the beta of the bipolar transistors. With this combination our only limitation in switching time is either the limiting speed of the CMOS driver or the f_t of the bipolar transistors. Emitter-followers never saturate, so we have no turn-OFF delay.

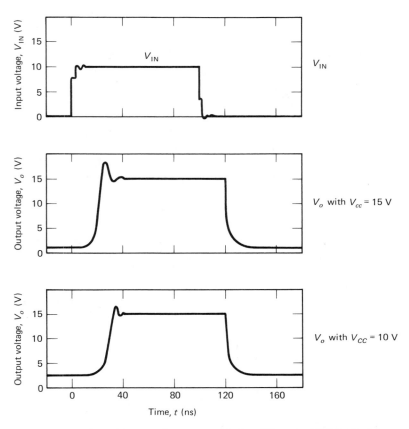

Figure 8-11 Switching performance of the Siliconix VN66AF driven by four paralleled CD4011 CMOS gates. (Courtesy of Siliconix incorporated.)

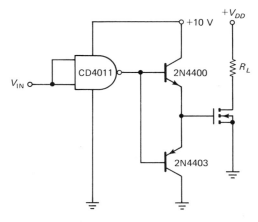

Figure 8-12 An emitter-follower increases the current source and sink to speed up switching time.

So far we have shown that it is pretty easy to drive logic-compatible power MOSFETs from high-voltage (±15 V) CMOS, but what about other combinations, for example, a high-threshold-power MOSFET? Some suppliers of high-voltage high-power power MOSFETs believe, and with good reason, that a high threshold maintains a level of protection from transient noise that is common to such applications as motor controllers and switchers. If we want to use one of these high-voltage power MOSFETs as a logic-controlled switch, we will have to rely on a driver capable of outputting a high voltage (15 V is always a good value) as well as being capable of providing a good dose of current since high-power power MOSFETs have rather respectable input capacities (for example, the IRF350[4] 400-V 11-A power HEXFET is rated at a nominal 3000-pF C_{iss} and the Siliconix VN4000A 400-V 6-A VMOS-FET is rated at a nominal 800-pF C_{iss}). There are other very suitable drivers that are equipped to handle both the higher threshold and the higher drive current, which we discuss next.

Driving the power MOSFET from RTL, DTL, or TTL poses a slight problem which we can easily overcome with little complication if our power MOSFET has a "logic-compatible" threshold. All of these logic elements operate at lower logic levels than does CMOS. Where we saw for our high-voltage CMOS drivers an output that could reach to the positive rail (+15 V), with these logic elements a high state would be in the neighborhood of +3.7 V and quite probably a bit lower. Although lying outside the scope of this book, we can, nonetheless, understand the reason for this low logic level by analyzing the typical TTL configuration shown in Fig. 8–13. From this figure we see that

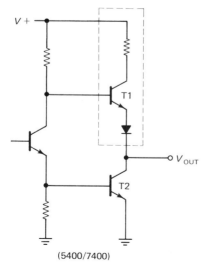

(5400/7400)

Figure 8-13 Basic TTL output configuration, clearly showing that V_{out} is at least two diode drops below 5 V.

[4] The IRF350 is also available as a 2N6768.

V_{out} is at least two diode drops below V^+, which for TTL is generally limited to +5 V.

For our power MOSFETs to be truly compatible with this logic, we need to raise the output voltage level to ensure that our FET is fully utilized. There are a variety of methods we can use. As we review the several methods, we must always keep in mind our need to both supply and sink possibly substantial amounts of current if we wish to switch our power MOSFET quickly. Equation (8.5) or our earlier equation (4.15) clearly identifies how gate current is proportional to C_{in} and inversely proportional to the switching time ∂t. We have not, at least in this chapter, discussed the *sink* requirements for current, but we must not forget that turning the power MOSFET ON requires us to charge C_{in} (Q_g), which then must be removed before the power MOS-FET turns OFF. The only way we can accomplish this latter task is to provide a current sink of low resistance capable of handling the discharge current in the required time.

Undoubtedly, the first "fix" a circuit designer would consider if standard TTL were being used would be an external pull-up resistor as shown in Fig. 8-14. This, however, is only a partial solution, for our switching speed is at the mercy of the RC time constant of the pull-up resistor and C_{in} of the power MOSFET. Furthermore, with the pull-up resistor going only to the +5-V rail, we are still unable fully to utilize the performance of the power MOSFET.

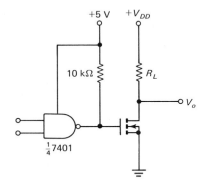

Figure 8-14 Driving a power MOS-FET with TTL, using an external pull-up resistor to enhance the gate-to-source voltage of the MOS-FET.

Open-collector TTL becomes the next logical choice, where with an external pull-up resistor we can tie to a higher voltage rail than the +5 V for the TTL. With this arrangement, as shown in Fig. 8-15, we can now fully utilize our power MOSFET by riding the gate voltage right to the upper rail. Again, as before, our turn-ON time is limited by the RC time constant of the pull-up resistor and C_{in}.

Open-collector TTL does offer us an excellent current sink to turn our power MOSFET OFF in a hurry. Our problem is in turning it ON

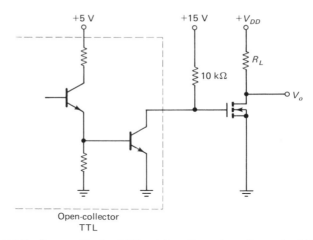

Figure 8-15 Open-collector TTL with pull-up that allows the MOSFET gate voltage to rise to +15 V, thus ensuring a full ON condition.

with equal speed.   We can see the reason for the excellent current sinking by noting the area in Fig. 8–13 omitted for open-collector TTL, transistor T_2 providing a low-resistance path to ground. Speeding open-collector TTL with a pull-up is accomplished by using what is known as a totem-pole emitter-coupled driver as shown in Fig. 8–16. Here we have added a high beta *npn* bipolar transistor and a diode. Together they offer us both a fast charge and discharge cycle to optimize the switching performance of our power MOSFET. Operationally, we can trace the performance as follows. With T_2 ON the base of our *npn* totem-pole bipolar transistor is tied *low*, ensuring that it is OFF. The gate of the power MOSFET is likewise OFF, as it is effectively only two diode drops above ground (T_2 and the diode), which, in turn, keeps the FET OFF. Now we turn T_2 OFF and the output of the TTL rises toward V^+ and the *npn* bipolar transistor turns ON, dumping nearly unlimited current into C_{in}. The diode, meanwhile, is effectively OFF, as the base and emitter of the *npn* bipolar transistor are at nearly the same potential. To appreciate more fully the effectiveness of this high-speed TTL, we need only see a typical waveform exhibited in Fig. 8–17. The VN66AF is a Siliconix power VMOSFET encapsulated TO-202 plastic package typically with 40-pF C_{in}.

8.6.2 Using Logic to Drive the Source-Follower Power MOSFET

We have now come to the seemingly difficult job of how to keep the gate of our power MOSFET at least 10 V above the source voltage, *irrespective of the source voltage*. This is an easy task if we have another voltage source handy that is already higher than our drain

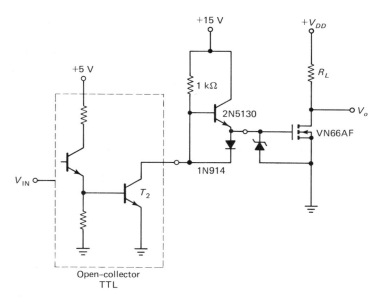

Figure 8-16 An external totem pole increases the switching time by offering nearly unlimited current to charge C_{in} of the power MOSFET. (Courtesy of Siliconix incorporated.)

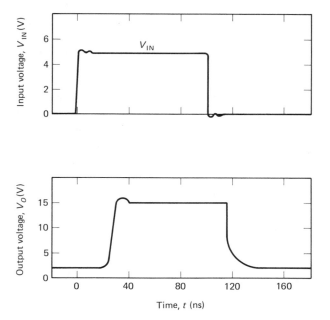

Figure 8-17 Switching performance of the Siliconix VN66AF when driven by open-collector TTL with totem pole (Fig. 8-16). Most of the delay is developed within the TTL driver and *not in the power MOS-FET*. (Courtesy of Siliconix incorporated.)

voltage by the necessary 10 or so volts. Generally, we are not so fortunate. But we can provide such a voltage "artificially" by using the
bootstrap technique, which should ring a familiar note, as we discussed
the bootstrap in some detail in Chap. 5. In that chapter our emphasis
was slightly different, for we were driving a totem-pole arrangement of
power MOSFETs. We will be considering that, too, in this chapter
when we discuss high-speed line drivers, but now let us concentrate on
driving the source-follower.

The basic source-follower with bootstrap is shown in Fig. 8–18.
We should not limit our thinking with regard to source-followers, as
we could just as easily substitute a load in place of the source resistor.
The action and the need for the bootstrap remains the same. For a
refresher as well as for convenience, we will review how the bootstrap
works. Let us start with our logic drive in the *low* state, which effectively grounds the gate of the power MOSFET and, of course, turns the
MOSFET OFF. The capacitor C is then charged through D_1 to the
supply voltage less any drop caused by the source resistor (or load).
The 10-kΩ resistor plays no role—yet. When our logic goes *high*, we
turn ON the MOSFET with positive gate voltage and as drain current
begins, the voltage across the load rises. Now, if we did not have this
bootstrap, we would be in trouble as soon as the current began, as the
rising source voltage would soon force a decrease in the effective gate-
to-source voltage which would work to shut down our MOSFET. We
obviously do not want this to happen. With the circuit shown in this
figure, as the source voltage rises, the voltage at the gate rises as though
there were a voltage source *in series with the source voltage*. If the
capacitor C is sufficiently large, we will be able to maintain the power
MOSFET ON for a few seconds. It is important that we recognize that
a bootstrap is good only for repetitive drive situations and not for
steady-state applications. Once the capacitor loses its charge, the gate
voltage drops and unless we recharge the capacitor by first shutting

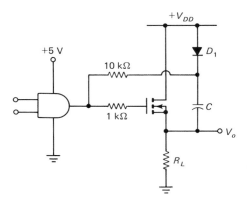

Figure 8-18 Driving a source follower from conventional TTL using
a bootstrap.

down the circuit, we will find ourselves thoroughly frustrated. D_1 should have a high reverse leakage resistance to help maintain the charge on the capacitor for as long as possible. A capacitor value of at least 10 times C_{iss} (there is no Miller effect in source-followers) is a good rule for most applications. If the load is inductive, we must remember to offer protection with a shunting diode (anode at ground).

An alternative method of driving a grounded load from logic is to use a *p*-channel enhancement-mode power MOSFET driven from CMOS. In the circuit shown in Fig. 8-19 we have a *p*-channel power MOSFET "upside down," with its source terminal tied to the positive voltage rail and the CMOS, in effect, working in reverse. When our logic output goes *low*, the power MOSFET turns ON, and when it goes *high*, the circuit shuts down.

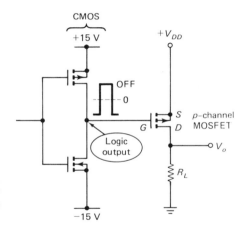

Figure 8-19 Driving a *p*-channel power MOSFET source follower from high-voltage CMOS avoids the need for bootstrapping.

8.7 USING THE POWER MOSFET IN HIGH-SPEED LINE DRIVERS

When we think of line drivers we tend to think of the problems of driving a highly capacitive coaxial cable. The problems of charge transfer that we discussed in Sec. 8.4 also occur when we wish to input a burst of energy into a highly capacitive coaxial line. We can rewrite Eq. (8.3) to agree more with our problem:

$$W = \tfrac{1}{2}(\Delta Q)\Delta V \qquad \text{watt-seconds} \qquad (8.8)$$

where ΔQ is the charge we have to "dump" into the coaxial line and ΔV is the voltage change that we wish to transmit along the line. If this information that we are trying to pass through this line is digital logic, we must be able not only to dump a heavy charge into the line but we have got to *remove* it with equal speed. Many of us are probably too familiar with the classic waveform comparison of an entering

waveform to that of an exiting waveform on a highly capacitive transmission line.

Since we not only have to source current into a current-hungry
load (a capacitor begins its charge cycle starting as a short circuit), but
we have to sink the same current with the same speed to discharge the
capacitor. We have already proven that it takes a power MOSFET to
source, or drive, the necessary current at the speeds we prefer to
operate, so we need a power MOSFET to sink the current quickly.
Consequently, we can design a high-speed line driver using the totem-
pole arrangement of power MOSFETs as shown in Fig. 8-20. Here we
have a pair of n-channel power MOSFETs in a circuit arrangement,
where the lower one operates in the conventional common-source mode
and the upper one operates as a source-follower.

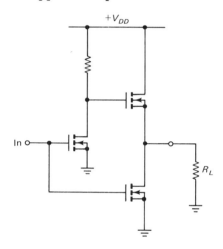

Figure 8-20 Basic totem-pole line
driver.

In Sec. 8.6 we learned how to drive both the common-source
MOSFET and the more difficult source-follower. Now we are faced
with a need to drive both from the same logic, not simultaneously of
course, but first we must source current into the line and then, at the
conclusion of the input pulse, to sink current from the transmission
line. Driving the common-source power MOSFET is both simple and
direct. To drive the source-follower we use the bootstrap circuit with
the phase-reversal power MOSFET input in shunt with the input to the
common-source power MOSFET as we have in Fig. 8-21. Because of
the very high input impedances offered by the gates, there is practically
no shunting effect (loading) to the driver. This bootstrap arrangement
looks different from the one we saw earlier in Fig. 8-18, and may
require a brief explanation. We already know that to turn a power
MOSFET ON fast requires sufficient gate current to charge the input
capacitor quickly. In fact, the faster we want to switch, the more

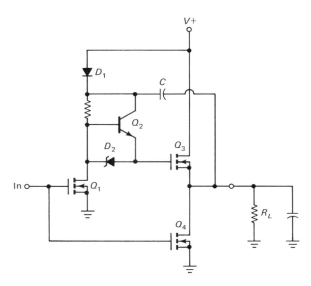

Figure 8-21 High-speed line driver with bootstrap.   (Courtesy of Siliconix incorporated.)

current is required.   This we saw from Eq. (8.5), but to make this equation more manageable, we can derive from Eqs. (8.4) and (8.5) the equation

$$I_g = C\,\frac{E}{t} \qquad \text{amperes} \qquad (8.9)$$

where E is the voltage we need to reach to turn the MOSFET ON and t is the time in seconds. For our illustration we might assume E to be 15 V, which will assure us of saturated ON operation and see how much gate current we would need to switch in 10 ns. Let the MOSFET have 50 pF input capacity. Our answer is 75 mA. It is pretty obvious to us that we cannot source that much current using the circuit shown in Fig. 8–18, and neither could we sink that much current in time to provide a fast shutdown. The solution is the circuit in Fig. 8–21 using a totem-pole bootstrap arrangement. Here is how it works. We have substituted an emitter-coupled bipolar transistor for the bleeder resistor. When we hit the gate of Q_1 with a positive pulse, Q_1 turns ON and the base of Q_2 is pulled down and diode D_2 clamps the gate of Q_3 to ground through the low $R_{DS(\text{on})}$ of Q_1, which guarantees that neither Q_2 nor Q_3 are ON. Q_4, of course, is ON by virtue of the positive pulse applied to its gate. Any charge on the gate of Q_3 is quickly discharged through D_2 and Q_1. Q_2 acts as an emitter-follower to increase the peak gate current to Q_3 when Q_1 is OFF. The rest of the circuit operates identically to our earlier description of Fig. 8–18.

We can still speed up the switching of our totem-pole line driver by using high-speed components such as Schottky diodes for D_1 and D_2 and taking a few necessary precautions such as placing a ferrite bead on each gate lead to reduce the possibilities of parasitic oscillations or undue ringing. We can speed up the action of Q_1 by using a gate speed-up circuit shown in Fig. 8-22, which provides a rather significant improvement in switching time. It does, of course, force us to increase the gate drive voltage. Operationally, the capacitor appears as a short circuit across the resistor which allows our inrush of current to charge C_{in}. Improvement comes when we switch OFF. Now the fully charged capacitor is now *in series with an oppositely charged* C_{in}. If our capacitor is many times larger than C_{in}, it will quickly force C_{in} to discharge, thus turning OFF our power MOSFET. We can get a better appreciation of how significant this speed-up circuit is by studying Figs. 8-6 through 8-8, which describe the improvement achieved with an International Rectifier HEXFET IRF330. The reason for the apparent lack of improvement at low drain currents stems principally from the RC_{oss} time constant. The lower our drain current, the higher the load impedance.

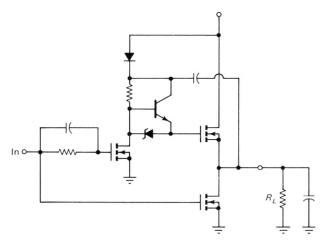

Figure 8-22 High-speed line driver with both bootstrap and input speed-up network.

An interesting circuit using a complementary pair of power MOS-FETs is shown in Fig. 8-23. We can visualize several possible driver circuits other than what we show here, but TTL is probably the best. There are several things we need to note in this circuit. First and undoubtedly the most obvious is that we no longer need a bootstrap to turn ON the uppermost *p*-channel MOSFET. Second, we need to observe that the *p*-channel MOSFET has its source tied to the positive

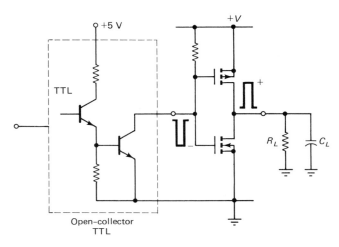

Figure 8-23   TTL-compatible complementary-pair MOSFET line driver.
Note that no bootstrap is necessary since the top MOSFET is a p-channel.

rail.   Remembering that to turn ON a p-channel enhancement-mode
power MOSFET requires that we pull the gate negative with respect to
the source makes this unusual circuit arrangement clear.   To turn ON
the p-channel MOSFET we merely clamp the gate to ground.   Clamping
the n-channel MOSFET gate to ground ensures that it is OFF.   Similarly, pulling both gates to the positive rail ensures that the p-channel
MOSFET stays OFF and the n-channel MOSFET turns ON.

There are, however, some basic problems with this circuit which,
depending upon the application, may require some careful examination.
A p-channel MOSFET has about two-thirds the mobility of an n-chan-
nel MOSFET, so to make a complement we generally match $R_{DS(on)}$,
which means, among other things, that the various interelectrode capacitances are far from equal.   The p-channel capacitances will be larger
than those of the n-channel MOSFET for equal $R_{DS(on)}$ and/or drain
current I_D.   Another obvious disadvantage is that we have both gates
tied together, which effectively places the input capacitance of both
power MOSFETs in parallel.   Regardless of which power MOSFET we
wish to turn ON, we are bound to the task of charging both gate
capacitances.   Furthermore, since both power MOSFETs are in common-
source, the Miller effect makes our total C_{in} somewhat painful, especially if we wish to switch fast.   Although there has been much discussion as to the merits of complementary pairs, there is little merit in
this application if high-speed switching is our goal.

We should also be aware that even if we are switching into a pure
resistive load, because of the input charging ramp that inevitably
occurs, for a portion of the switching time both MOSFETs will be

partially ON, thus effectively shorting the rails. The resulting current spike will affect not only our output pulse waveform, but it will also affect the dissipation of our MOSFETs.

8.8 INTERFACING THE POWER MOSFET WITH THE MICROPROCESSOR

We purposely left this until last, as by now we should have a good idea of how to interface the power MOSFET to nearly all types of logic, with the exception perhaps of direct-interfacing high-voltage MOSFETs to 4000-series CMOS logic, where the voltages are too low to switch the power MOSFETs effectively. Interfacing to a microprocessor poses no major problem. We have the option to go through the input/output, or depending upon our needs, we can interface directly to any CMOS microprocessor if the logic levels are such as to afford useful current handling by our power MOSFETs.

8.9 BENEFITS AND PROBLEMS IN USING POWER MOSFETS

We have seen in this chapter that the n-channel enhancement-mode power MOSFET is an outstanding candidate for use in logic-driven applications, where substantial power control is required from digital control. The benefits are obvious. There are, however, some problems that should be considered. As industry continues to offer power MOS-FETs with increasingly higher voltage and current ratings, we should expect to see rising threshold voltages, which will be justified on the grounds that a higher threshold voltage will reduce sensitivity to noise and transients that might cause false triggering. If this event occurs, we will find these devices incompatible with most logic families except for high-voltage CMOS. On the other hand, with logic-compatible power MOSFETs we find a temperature-related problem that under severe environments may see our power MOSFET more susceptible to false triggering because of a downward drift in threshold with increasing temperature. Nevertheless, irrespective of these potential hazards, the power MOSFET is the most useful interface between logic and analog that we have seen for some time.

REFERENCES

EVANS, ARTHUR, and DAVID HOFFMAN, "Dynamic Input Characteristics of a VMOS Power Switch," *Siliconix Application Note AN79-3*. Santa Clara, Calif.: Siliconix incorporated, 1979.

PELLY, BRIAN R., "Applying International Rectifier's Power MOSFETs," *International Rectifier Application Note AN-930*, El Segundo, Calif.: International Rectifier, 1980.

nine

Using the Power FET as a Switch

9.1 INTRODUCTION

We might think that this chapter title is inappropriate, since most of the applications that we have studied up to now have identified the power FET as a nearly-perfect switch. Indeed, our criticism would be justified if we narrowed our perspective of switching to only that which we have covered in earlier chapters. If this seems bothersome, we should again pause to reflect upon and to compare the switching we have already studied with what might be commonly understood as switching by a more general audience. If we were to define what *switching* means in both the industrial and consumer markets, we would then appreciate that these earlier chapters only touched upon a narrow segment of switching. In Chap. 6, on motor control, as well as in Chap. 5, on switching power supplies, we were interested primarily in PWM switching; in Chap. 8, on using power FETs in logic control, our focus was on digital interface applications. Industrial and consumer markets are, for the most part, principally concerned about the control of *analog signals*.

Analog signal processing may be divided into two broad categories:

1. Handling low frequency ac signals
2. Handling high-frequency ac signals

We should be aware that, unlike what we have studied in previous chapters, these switches do not necessarily need to be logic-compatible. Our intent in this chapter will be to examine the usefulness and the practicality of using power FETs in various switching circuits that need to remain either ON or OFF for indeterminant periods of time. We must remember that our earlier discussion of power FETs as switches was entirely devoted to pulse-actuated switching, either by some form of logic or by PWM. The switch characteristics that we will be concerned about will include $R_{DS(on)}$ and how it behaves as the analog signal voltage varies, charge injection (which differs from our previous definition of charge transfer), insertion loss, and OFF isolation.

There is really nothing particularly spectacular about using transistors as analog switches. Driver-gates using J-FETs as well as MOS have been around since the mid 1960s and have competed quite successfully against the mechanical switch or relay. There have, of course, been arguments both pro and con for *both* the mechanical *and* the solid-state (transistor) switch! We can summarize these arguments as follows.

9.2 COMPARISON OF MECHANICAL AND FET SWITCHES

Because of the wide diversity of both mechanical and semiconductor switches available today we must qualify the type of switches we wish to compare. It would be preposterous of us to compare a relay used to control a diesel-electric locomotive with a solid-state analog switch found in a computer. It might not even be fair of us to compare the mechanical switch we use to turn on our TV with that used in a video camera at the studio. Any comparison must be viewed according to the application.

The most obvious comparison between a mechanical switch, for example, a relay, and a FET switch, such as a J-FET or MOSFET, would be the difference in ON resistance. Any *new* mechanical relay would exhibit a very low ON resistance, whereas the greater majority of existing J-FET or CMOS switches (commonly referred to as *gates*) have ON resistances hundreds or even thousands of times greater. Based on the extraordinarily low ON resistance of the relay we might erroneously disregard the solid-state FET switch. ON resistance cannot be our sole determinant; the application must support our decision. That diesel-electric locomotive would never use a solid-state FET switch. Applications requiring the switching of thousands of amperes with standoff voltages also in the thousands will probably rely on mechanical switches forever or on SCRs. On the other hand, we have probably seen small relays in modern electronic equipment. Here we may be on safer ground to make logical comparisons; but again, of course, we must be aware of the circumstances, or, in other words, the application. If, in this example (the computer) we needed to control high currents we would obviously need a very low resistance switch, and a relay might be our most obvious choice. On the other hand perhaps we are switching logic or maybe even analog signals on a high-impedance line. Now an ON resistance in milliohms would be a senseless luxury. Would it make any difference if a NAND gate outputted to $1m\Omega$ or to $30~\Omega$ in series with the next logic function? Of course not. Now we have made our first comparison, and there are more to come.

We really need not compare the switching speed of a mechanical switch with that of a FET switch. No comparison is necessary. Mechanical switches might switch in milliseconds; FETs, with no minority-carrier storage time, switch in nanoseconds. Of course, if our application does not require high-speed switching, we need to make our decision on other comparisons.

Standoff voltage might be another area for comparison. Mechanical switches can be built to withstand practically any voltage. The FET switch is limited to the breakdown voltage rating. Again the application will determine our selection.

If our switch must be controlled through some degree of involvement with electronic systems, for example, needing to be driven direct from logic, we might find a mechanical switch somewhat more of a burden, whereas the FET switch would offer us the opportunity for direct logic control.

Moving parts are an anathema to high reliability and long life, so the relay must acquiesce to the FET switch. MTBF (*mean-time-between-failure*) for FETs is in the tens of billions of operations, whereas for mechanical switch or relay—well, there is simply no comparison.

9.3 THE SMALL-SIGNAL FET VERSUS THE POWER FET AS A SWITCH

The contemporary analog switch, or as it is often called, the analog gate, in its simplist form as a *single-pole single-throw* (SPST) switch, can be one of three possible types: J-FET, N-/P-MOS, or CMOS, as shown in Fig. 9–1. Each of these types has numerous advantages and disadvantages, depending, of course, on the application. If our desire was for the lowest ON resistance, our choice would be the J-FET, where resistances ranging from 10 to 30 Ω are typical. The major disadvantage of the J-FET switch which has popularized the various MOS switches (P-MOS and CMOS) is its limited analog signal voltage

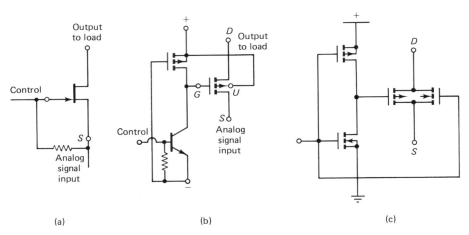

Figure 9-1 Three basic SPST solid-state analog switches: (a) J-FET;
(b) PMOS; (c) CMOS. (Courtesy of Siliconix incorporated.)

capability. Using a J-FET as an analog switch with CMOS drivers in,
for example, a logic system operating with 15-V rails, our J-FET
switch would be limited to switching analog signals not much greater
than ±10 V. A CMOS switch could switch to the rails (±15 V) but with
a much higher $R_{DS(on)}$. An additional advantage of the J-FET switch
is its low distortion, resulting from negligible *resistance modulation*, a
characteristic more difficult to control with either a P-MOS or CMOS
switch.

A common fault for all these contemporary small-signal analog
switches is their power-handling capability. With the best J-FET
switches sporting 5- to 10-Ω ON resistances, it is unlikely that they
could handle more than a few hundred milliamperes, at best. Furthermore, their standoff voltage seldom reaches beyond 40 V. For logic
switching, sample and hold, some video switching, and a variety of
instrumentation and multiplexing applications, these small-signal FET
switches have wide application. But there remains a wide field of
potential applications where we must have both ultralow ON resistance
and high-standoff-voltage capability, and these devices simply will not
do.

The SIT in its present state has several unfortunate characteristics
that may prevent its use in many analog switching applications. Chief
among these are the high pinch-off voltage required to turn the SIT
OFF, and the unusually high parasitic drain-to-gate capacitance. An
intriguing characteristic of the SIT, which we touched upon briefly in
Chap. 2, is the effect that positive gate bias has on performance. In Fig.
2–18, we witnessed a depletion-mode J-FET when negative gate bias is
applied and what appears as an *npn* bipolar transistor with positive gate

(base?) bias. Although outside the scope of this book, its operation as an analog switch might be an interesting project for study.

The power MOSFET overcomes most of the problems that have been associated with small-signal analog switches, offering extremely low ON resistance and high standoff voltage capability and with low to moderate interelectrode capacity, especially between gate and drain. This ultralow $R_{DS(on)}$ characteristic of the power MOSFET allows us to switch reasonably high currents with little dissipative power loss and, for any analog signal switching applications introduces very little distortion that otherwise might have been caused by *resistance modulation.*

The power MOSFET, however, is not the ubiquitous analog signal gate that we might have wished it to be simply because of that so-called *beneficial* parasitic *p-n* diode that exists between the *p* channel and the *n*— drain–drift region that we discussed in some detail in Chap. 4 and in Fig. 4–20. The equivalent circuit for a power MOSFET gate in both the ON and OFF states is shown in Fig. 9-2. The problem we have is

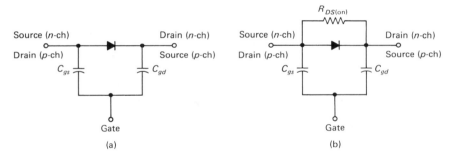

Figure 9-2   Simple equivalent circuit for a power MOSFET in the OFF state (a) and in the ON state (b).

obvious. If, in the OFF state our analog signal voltage exceeds the "turn-on" voltage of this parasitic drain–source diode, we have conduction rather than continued isolation as we might wish for the OFF state. Our power MOSFET gate becomes, in effect, a half-wave rectifier, and the resulting output waveform looks much like that presented in Fig. 9–3! Really not quite what we want. The solution, fortunately,

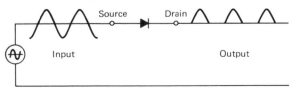

Figure 9-3   Half-wave rectification resulting from using a power MOSFET as a series gate for analog signal switching.

is rather obvious, but we reserve that until a bit later, for now we should turn our attention to the vexing problem of analog signal distortion, which we claim, the power MOSFET helps to alleviate.

9.4 REDUCING DISTORTION CAUSED BY RESISTANCE MODULATION

Distortion of our analog signal passing through our gate occurs if the ON resistance ($R_{DS(on)}$) varies with the analog voltage in anything but a linear relationship. What is meant by that is simply shown by the following Ohm's law equation:

$$R_{DS} = \frac{V_{DS}}{I_D} \tag{9.1}$$

which, for our power MOSFET, can be rewritten from Eq. (4.11) as

$$R_{DS} \simeq \left(g_m - \frac{|V_{DS}|}{2} \right)^{-1} \tag{9.2}$$

Where g_m is [repeating Eq. (4.11)]

$$g_m = \frac{\mu \epsilon_{ox} W (V_G - V_{TH})}{L T_{ox}}$$

It now becomes patently clear that $R_{DS(on)}$ is dependent upon gate-to-source voltage V_G. If we were to place a power MOSFET as a series gate, such as shown in Fig. 9-4, we could easily calculate the insertion loss at low frequencies both with and without the effects of resistance modulation.

$$\% \text{ error} = \begin{cases} \dfrac{-100}{1 + \dfrac{R_L}{R_{DS}}} & \text{(without resistance modulation)} \quad (9.3) \\[20pt] \dfrac{-100}{1 + \dfrac{R_L}{R_{DS} + \Delta R_{DS}}} & \text{(with resistance modulation)} \quad (9.4) \end{cases}$$

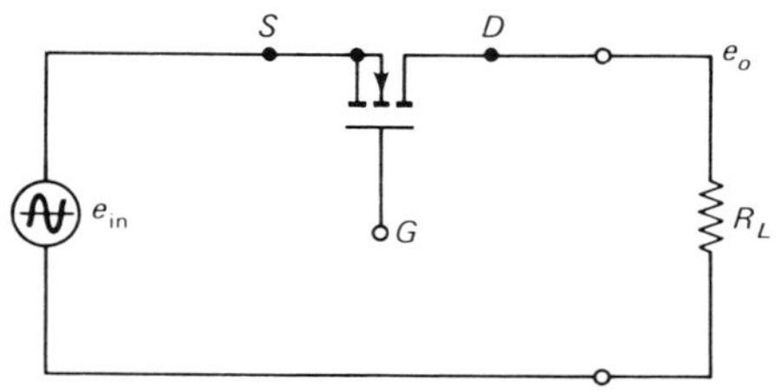

Figure 9-4 Power MOSFET as a simple gate.

We can go one step further and determine the attenuation:

$$\text{I.L.} = 20 \log \frac{\% \text{ error}}{10} \quad \text{dB} \tag{9.5}$$

At high frequencies the problem becomes more complicated because of the parasitic capacitances between source and gate as well as drain and gate as we saw in Fig. 9–2, which forms a capacitive voltage divider. We will not analyze the high-frequency distortion due to parasitic capacitance feedthrough, simply because it ends up as a futile exercise. Using power MOSFETs as high-frequency gates is both feasible and satisfying, but because of the many parasitic elements involved in the actual construction, performance is best determined by measurement.

Equations (9.3) and (9.4) clearly indicate that as we increase the ratio $R_L/R_{DS(\text{on})}$, the error decreases proportionally. This is one reason why using a very low $R_{DS(\text{on})}$ power MOSFET makes such a good analog switch. All that now remains for us is to deduce the deleterious effects of changing V_G on $R_{DS(\text{on})}$ and to resolve how to overcome the half-wave rectification problem that occurs in the OFF state.

Power FETs also function as *voltage-controlled resistors* (VCRs) by varying the gate-to-source voltage. Again, we must be careful that distortion caused by resistance modulation does not upset our application. The solutions that we shall discuss for analog switches also applies for VCRs.

9.5 COPING WITH CHARGE-TRANSFER EFFECTS

In our study of transient analysis in Chap. 8 we saw how parasitic capacitances affected our drive power. What we are about to consider here will be a close parallel.

In any solid-state switch used as an analog gate, the voltage transients that appear at the gate of the FET (J-FET, N-/P-MOS, or CMOS) when it is turned either ON or OFF will be coupled through the parasitic capacitances C_{gs} and C_{gd} into the signal path and appear as spikes, or *glitches*, the term most frequently used in industry. The magnitude of these spikes is dependent upon several factors: (1) the amplitude of the driver output voltage used to bias the power MOSFET gate, (2) the speed of the output voltage pulse, (3) the size of the parasitic capacitances coupling between the gate and analog signal channel, (4) the load impedance (as well as the source impedance to a lesser extent) and (5) the impedance of the driver. Some of these factors are readily identifiable in Fig. 9–5, where we have equivalent circuits for the power MOSFET in both the ON state and the OFF state.

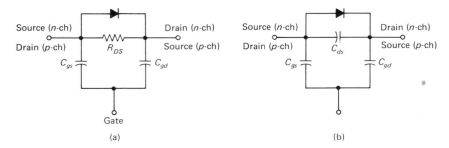

Figure 9-5 Equivalent circuits for the power MOSFET in the ON state (a) and in the OFF state (b).

The fact that spikes, or glitches, occur in the output of our MOS-FET switch should come as no surprise, and the five factors, which we have just identified, regulating the magnitude of these spikes should appear entirely reasonable. Let us consider what we have got and why spikes should be expected. Our power MOSFET switch is a *three-terminal device*: source, drain, and gate. In our study of transient analysis in Chap. 5, we noted that to bring the gate potential up and beyond the threshold voltage required a finite charge, which could be identified as a gate current. If, indeed, our power MOSFET switch is a three-terminal device, we must presume that the charging current (gate current) needed to charge C_{in} *also flows into the signal path.* Where else can it go? The charging current must flow as shown in Fig. 9-6, developing a brief voltage spike across R_L that resembles a capacitive charging time constant.

Before we become unnecessarily alarmed by the possibility of

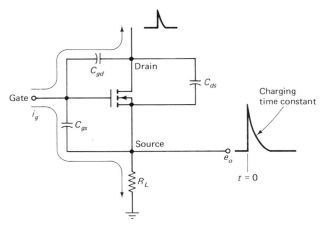

Figure 9-6 Flow of charging current that results in a "glitch" in the output waveform.

spikes appearing on the output of our analog signal gate, we must recognize that if our application called for using a power MOSFET instead of a more common small-signal analog switch, we are either handling substantial power or holding off substantial voltage. In the former situation our source and load impedances would be quite low, and in the latter case the voltage would be sufficiently high as to make the spike voltage a small fraction of the total. Rewriting Eq. (8.4), we can determine the charge transferred:

$$Q = CV \tag{9.6}$$

Calculating the charge transfer or the magnitude of the spike is no easy task, and if attempted we would find the task disagreeable because of the many indeterminants involved based on the interrelationships that we itemized earlier that influence the magnitude of the spikes. Our best solution is to build a test circuit and measure the amplitude of the spikes appearing on the analog signal output. Since these switching spikes occur only at turn-ON and only for the duration required to charge C_{in}, the longer the duration of our ON time, the more inconsequential they become.

9.6 OVERCOMING THE HALF-WAVE RECTIFICATION EFFECT

As we saw in Fig. 9–3, a switch standing in the OFF state and producing an output waveform is certainly *not* the switch we would want unless we were controlling dc rather than ac (analog voltages). We studied the phenomenon that gave rise to this parasitic diode in Chap. 4, where we were warned that although generally considered a beneficial parasitic for most applications, for switching it was not. Since every power MOSFET, irrespective of type—VDMOS, DMOS, or VMOS—has this *beneficial p-n* diode, it is crucial that we inhibit rectification if we wish to capitalize on the many features of power MOSFETs as analog switches.

The solution is as simple as it is obvious, as shown in Fig. 9–7, where we have two identical power MOSFETs *back to back*. The OFF leakage of the pair becomes the reverse diode leakage of either *p-n*

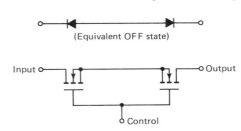

Figure 9-7 Basic circuit of a successful analog switch using power MOSFETs.

body-drain diode, which at low frequencies (under 1 MHz) can provide uncompensated insertion loss in excess of 30 dB, depending on the selection of power MOSFETs. The single fault of this scheme is, of course, the doubling of insertion loss when in the ON state. Although Fig. 9-7 shows a pair of *n*-channel MOSFETs, *p*-channel MOSFETs could have been used with equal success. We must again remember that *p*-channel MOSFETs exhibit higher $R_{DS(on)}$ than physically comparable *n*-channel MOSFETs.

A novel analog switch that might find use in protecting CMOS from a loss of supply power is the complementary analog switch shown in Fig. 9-8, where we must take special pains to have the drain–body

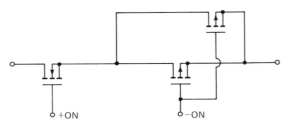

Figure 9-8   Circuit showing how combining *n*- and *p*-channel MOSFETs we can achieve a CMOS protection circuit. Should either rail voltage (+ or − 15 V) fail, the analog switch will turn OFF, interrupting the analog signal feeding the CMOS circuitry.

diodes *back to back*. In this figure we show a pair of *p*-channel MOS-FETs in parallel to emphasize the need of resistance-matching the $R_{DS(on)}$ to the *n*-channel MOSFET for optimum performance. The *n*-channel would require a positive gate voltage to turn ON, whereas the *p*-channel requires a negative voltage. Only if both a positive and a negative voltage are applied simultaneously would the analog switch conduct.

Another novel analog switch using only one power MOSFET is shown in Fig. 9-9. Because of the bridge arrangement of *p-n* diodes,

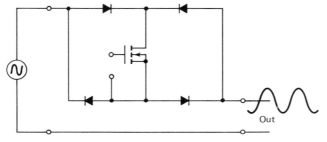

Figure 9-9   Simple MOSFET analog switch using a single MOSFET and four diodes. (Courtesy of Siliconix incorporated.)

there is no half-wave rectification; however, we do suffer in that the forward conduction is inhibited by $R_{DS(on)}$ plus *two* diode drops. The advantage of this bridge arrangement would be economical if we could purchase four diodes for less than the cost of one power MOSFET.

9.7 STABILIZING THE GATE VOLTAGE V_G

We noted earlier that signal distortion caused by resistance modulation was the direct result of a varying gate-to-source voltage in the presence of an analog signal. We can easily visualize the cause of this variation of the gate-to-source voltage by examining Fig. 9–10 and assuming, for our illustration, that we have a dc bias of +10 V on the gate to allow con- duction of a 5-V peak-to-peak analog (ac) voltage. It should be evident that although we have placed a +10-V dc bias on the gate of our series- connected MOSFET switch, the instantaneous ac gate-to-source voltage can swing from +7.5 V to +12.5 V. We have mentioned both in this and in earlier chapters that power MOSFETs can be used as voltage-variable resistors (VCRs) merely by changing the gate-to-source voltage, which, in turn, modulates the inversion layer and thus changes the channel resistance. The effect of changing V_{GS} on $R_{DS(on)}$ is shown in Fig. 9–10(b). Returning to our somewhat gross illustration we discover that resistance modulation becomes a function of the peak-to-peak analog signal amplitude if the gate voltage is fixed. It should now be obvious to us that to minimize resistance modulation, we must provide some suitable means of allowing the gate potential to follow the analog signal voltage. In other words, we must modulate the dc gate bias with the analog signal voltage.

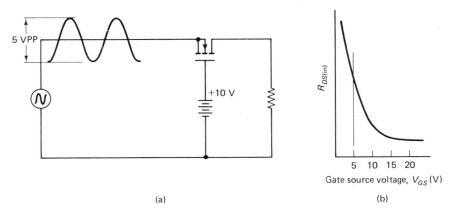

(a)   (b)

Figure 9-10   Varying the analog signal voltage can result in modulating the ON-resistance.

The problem of resistance modulation is certainly not unique to the power MOSFET analog switch. All analog switches using FETs suffer to some degree (except for the CMOS analog switch) if no preventive measures are taken in their design. One of the most effective means of modulating the gate of the FET switch is simply to tie it to the analog signal when the switch is in the ON state, as shown in Fig. 9-11 for the popular Siliconix DG181-series analog switches. To operate this circuit we drive the *pnp* bipolar transistor's base positive, turning it OFF while simultaneously driving the gate of the PMOS, T1, negative, turning it ON, which effectively ties the gate of the J-FET, T2, directly to its source, thus ensuring it to remain hard ON irrespective of the analog signal voltage peak-to-peak excursions. If we do not have a driver with complementary outputs (V_{OUT} and $V_{\overline{OUT}}$), we can reduce the effects of resistance modulation by tying a low-value resistor between the gate and source of the FET switch. What we have shown by this illustration is exactly the method we can use to reduce the resistance modulation of our power MOSFET analog switch, which we now address.

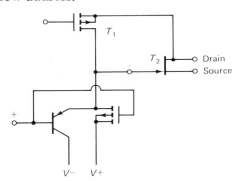

Figure 9-11 Output state of a typical J-FET SPST analog switch. (© 1976 Siliconix inc., reprinted with permission from *Analog Switches and Their Applications.*)

9.8 LINEARIZING THE POWER MOSFET ANALOG SWITCH

Although we may be convinced by this illustration, a careful reader might quickly note the fundamental difference between the operation of a J-FET analog switch and an enhancement-mode MOSFET switch. To turn the J-FET ON requires only that its gate-to-source voltage be zero, whereas for our enhancement-mode MOSFET we must raise the gate-to-source voltage sufficiently beyond threshold to ensure a low $R_{DS(on)}$. Consequently, our scheme to modulate the gate will be somewhat more complicated than that offered in the example shown in Fig. 9–11.

We know that to prevent resistance modulation the gate potential must be modulated in phase by the analog signal and that the magnitude of this modulation must match the magnitude of the analog signal. We also recognize that the gate voltage must be sufficiently higher than the source voltage (more positive for an n-channel MOSFET) to provide a low $R_{DS(on)}$. One very successful solution is for us to use a unity-gain level shifter to sample the analog signal voltage and bias the MOSFETs gate. The easiest solution is to use a noninverting unity-gain operational amplifier (op amp) with the output bootstrapped upward with a 10- to 12-V zener diode, as shown in Fig. 9–12. If we bridge the bootstrap circuit with a moderate-size capacitor, as shown, we ensure that the gate voltage will follow the analog signal with less delay,

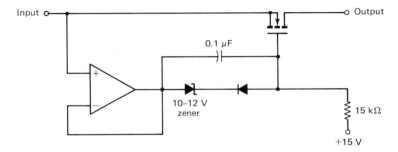

Figure 9-12 Buffering and bootstrapping the gate reduces resistance modulation and distortion. (Reprinted with permission from *Electronic Design*, Vol. 25, No. 15 © Hayden Publishing Company Inc., 1977.)

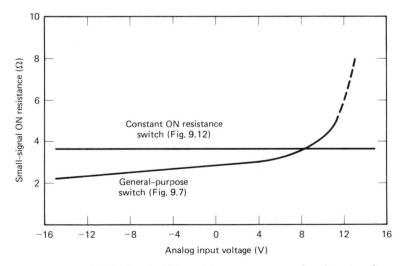

Figure 9-13 Small-signal ON resistance versus analog input voltage. (Reprinted with permission from *Electronic Design*, Vol. 25, No. 15 © Hayden Publishing Company Inc., 1977.)

thus assuring extremely low resistance modulation and greatly reduced distortion.

Now that we have settled the problem of resistance modulation, let us look at a few examples and compare their distortion performance both with and without our gate modulation bootstrap. Since we want a true analog switch, we will use a pair of MOSFETs in a circuit similar to that in Fig. 9-7. In Fig. 9-13 we see quite clearly that the problem of resistance modulation becomes gross at high analog signal levels. We might find it interesting to compare the swing of $R_{DS(on)}$ with that shown in Fig. 9-10(b). The resemblance is striking, as we might expect. The incredible improvement in total harmonic distortion (THD) between a nonbuffered MOSFET analog switch and a buffered analog switch, shown in Fig. 9-14, is very convincing proof that we should always employ some means to reduce the resistance modulation effect.

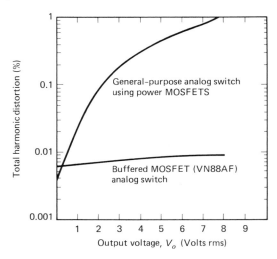

Figure 9-14 Total harmonic distortion vs. output voltage comparing a nonbuffered analog switch with a buffered analog switch. (Reprinted with permission from *Electronic Design*, Vol. 25, No. 15 © Hayden Publishing Company Inc., 1977.)

9.9 DRIVING THE POWER MOSFET ANALOG SWITCH

By now we all take for granted the fact that FETs operate with very little gate current unless called upon to switch very fast. In all but the most extraordinary circumstances, we would not expect to switch our power MOSFET analog switch fast, that is, in a few nanoseconds. As a consequence, all the gate control we need to turn our switch ON would be nothing more than a gate voltage sufficiently above V_{TH} to ensure a low $R_{DS(on)}$. This we can accomplish merely by our bootstrap circuit, whose original purpose was to reduce resistance modulation. Turning our analog switch OFF, however, requires that we pull the gates down below V_{TH}. Remembering that V_{TH} is relative to the source and, therefore, to the peak amplitude of the analog signal, we will need an external control capable of clamping the gates to a negative supply. Figure 9–15 provides us with the bare essentials for a low-distortion, low-ON-resistance analog switch.

Upon close examination of this basic circuit we discover that we could further improve the OFF-isolation if we used a double-pole single-throw (DPST) switch to crowbar the power MOSFET's source

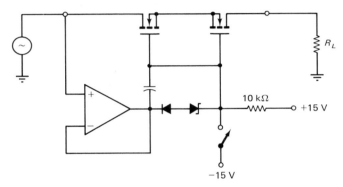

Figure 9-15   Basic low-distortion buffered analog switch.

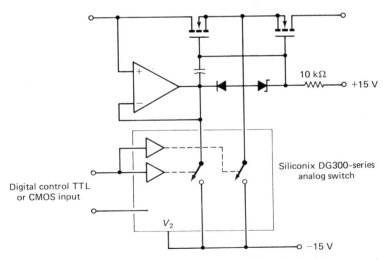

Figure 9-16   Fully buffered low-distortion, low-ON-resistance analog switch. (Reprinted with permission from *Electronic Design*, Vol. 25, No. 15 © Hayden Publishing Company Inc., 1977.)

terminals to the negative supply or to ground. Such a circuit would provide an extra measure of attentuation to our analog signal that otherwise might feed through the parasitic source–drain capacitances. The complete low-distortion, low-ON-resistance, general-purpose analog switch, fully TTL- or CMOS-compatible, is shown in Fig. 9-16. ON-resistance and distortion performance for this power MOSFET analog switch are offered in Fig. 9–17, where we see the improvement in both the ON-resistance and distortion using the unity-gain buffer and bootstrap circuit over what we would have had without it.

There are, of course, a variety of ways that we can drive our power MOSFET analog switch besides using a logic-compatible small-signal analog switch. We could use another power MOSFET, a bipolar transistor, a relay, or even a simple toggle switch for manual control.

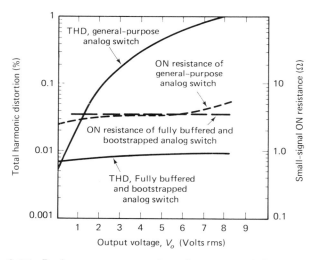

Figure 9-17 Performance comparison between a fully buffered and bootstrapped MOSFET analog switch and a general-purpose MOSFET analog switch.

9.10 SWITCHING POWER

As ON-resistances continue to drop and as packaging improves to the point where the thermal resistance equals the best of the commercially available power bipolar transistors, we will see increasing use of power MOSFETs in handling respectable amounts of both current and voltage in a variety of switching applications. Nevertheless, no matter how low our ON-resistance drops, we may at some time find a need for even lower resistance.

In Chap. 1, as we compared the power FET with the power

bipolar transistor, our attention was drawn to a fundamental advantage of a bulk semiconductor that if it had been the only advantage, it would have still made the power FET superior in many ways over the conventional power bipolar transistor. The FET is *thermally degenerate*; as current flow heats the chip, the bulk resistivity increases, which tends to act as a form of self-regulation to the current flow. We saw that this remarkable feature was principally responsible for the lack of current hogging, thermal runaway, and second breakdown. In our attempt to increase the power-handling capability of our power MOSFET analog switch, we find this thermal characteristic to our advantage, for we can parallel power MOSFETs endlessly without worry of current hogging. Furthermore, because of their high gate resistance, it takes no more effort to drive a pair than it does a dozen; provided, of course, that we keep our switching speed down. An additional benefit of paralleling power MOSFETs—or using low-ON-resistance power MOSFETs— is that distortion is a function of the ratio of $R_L/R_{DS(on)}$, as we saw earlier in Eq. (9.3), and with a low insertion loss [see Eq. (9.5)] and a high load impedance our problems of distortion become vanishingly small. If our standoff voltages are low, we could very effectively use some of the new ultralow $R_{DS(on)}$ power MOSFETs and get by without the need for the bootstrap linearity control. If we chose not to use the bootstrap, we would need to place a pull-up resistor from some positive rail (higher by at least +15 V than the peak positive analog voltage being switched) to the gate. Our turn-off logic controller (Fig. 9–16) would also control turn-ON.

So far we have considered high power to mean high current, but it could also mean high voltage, and if it did we would have a problem. If we reviewed several power MOSFET data sheets, we would be able to select from a wide variety of ON-resistances and breakdown voltages. Our first concern might *erroneously* be to find a balance between high breakdown voltage and low $R_{DS(on)}$ [mutually exclusive in light of Eq. (4.13)]. In trying to select both a low $R_{DS(on)}$ and a high BV_{DSS}, we might fail to observe the maximum allowable gate-to-source voltage specified on the data sheet. Our *first* concern should not be trying to find a balance between BV_{DSS} and $R_{DS(on)}$, but finding a power MOSFET with a high gate-to-source breakdown specification. To settle this clearly in our mind, let us consider an example.

Let us say that we want to switch a 120-V ac main. For this example we will not bother about the load or the possible distortion effects; they are immaterial to this illustration. Let us concentrate our attention on the matter of gate control. Earlier in this chapter we saw the need to bring the gate voltage well above the instantaneous source voltage. For most power MOSFETs a gate-to-source voltage of nominally +15 V is adequate to ensure a low $R_{DS(on)}$. To turn OFF our

MOSFET analog switch, we need to drop the gate-to-source voltage below threshold, V_{TH}. The problem that we want to emphasize in this illustration is our gate-to-source breakdown-voltage rating. Are there any power MOSFETs available capable of withstanding the potential gate-to-source voltage suggested by this example? Probably not. The reason should be clear. Threshold voltage is established, in part, by the oxide thickness between gate and channel, and the voltage breakdown is controlled partially by the same oxide. We would find it difficult or impossible to offer high gate-to-source breakdown and low threshold simultaneously. As we think about this, we must not forget that the channel is electrically tied to the source.

9.11 SWITCHING HIGH STANDOFF VOLTAGES

We closed the preceding section on a somewhat negative note, showing a difficulty in switching high *analog* voltages with power MOSFETs. Here we plan to show how we can safely switch high *standoff* voltages and maintain compliance with all safety codes. A sure-fire way to satisfy most safety codes is to provide complete isolation between the controller and the switch. We have such options if we use air-core transformers and optocouplers. Iron-core transformers will do provided that we can certify their insulation standoff voltage ratings.

Owing to the uniqueness of FETs in general, we are able to implement an isolated gate drive circuit with little complication since no gate current is required for moderate turn-ON speeds. The only complication, easily resolved, is that ac couples through transformers and optoisolators, whereas we need dc to activate the gates. The solution in either case is to couple ac and then rectify with the positive potential tied to the gates. The circuit shown in Fig. 9-18 provides a J-FET to clamp the gates to the sources for quick turn-OFF. Diode D_1 provides positive voltage to the gates, and diode D_2 places a negative voltage on the J-FET gate to keep it cut off until the ac signal is removed, signaling turn-OFF.

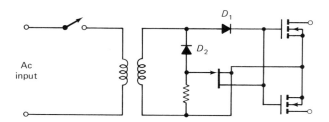

Figure 9-18   High-standoff-voltage analog switch using MOSFETs.

The ac signal needed to activate this transformer-isolated analog switch can be most anything, depending upon the transformer. An iron-core transformer could use subaudio through audio, whereas an air-core transformer might use upward of several megacycles.

An alternative to this transformer-coupled analog switch is the optocoupled circuit shown in Fig. 9-19, where a string of photovoltaic diodes, excited by the activated *light-emittting diode* (LED), generate sufficient dc voltage to turn the power MOSFETs ON. This circuit is especially useful since we are able to drive the LED, and hence the analog switch, directly from TTL.

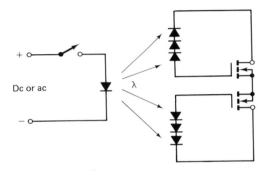

Figure 9-19   High-voltage standoff analog switch using photo-sensitive diodes and LEDs (Theta-J Corporation). (Reprinted with permission from *Electronic Design*, Vol. 28, No. 7 © Hayden Publishing Company Inc., 1980.)

9.12 THE POWER FET AS A VOLTAGE-CONTROLLED RESISTOR

If we were to inspect the saturation characteristics of a FET at low drain-to-source voltages, say under 5 V, we would discover that our FET's voltage and current relationships closely resemble those of a variable resistor. FETs make ideal voltage-controlled resistors *VCRs* in circuits requiring feedback control of resistance such as we might need in a Weinbridge oscillator or in voltage-controlled attentuators, and we can compare their performance with that of a conventional variable resistor in Fig. 9-20. In what is typically called the non-saturated region of a FET's output characteristics, we find that $R_{DS(on)}$ is under the direct control of the gate voltage V_{GS}. We saw that in Fig. 9-10(b) in our discussion of resistance modulation. Once the FET reaches *saturation*, the gate loses control of the ON resistance when operating under *dynamic* conditions, but under *static* conditions our gate maintains control. The current regulators that we described at the close of Chap. 5 were, in effect, VCRs operating in saturation and under essentially static conditions (where the drain-source voltage

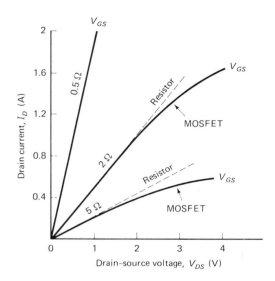

Figure 9-20 The power MOSFET behaves as a voltage-variable resistor with little distortion below saturation.

remains nearly constant). When operating in this fashion it is very important to have adequate heat sinking, since we have the worst of two worlds: high currents and high $R_{DS(on)}$.

VCRs suffer from distortion much the same as the analog switch if we use them in analog signal circuits. The cure is very similar to the cure for resistance modulation with one notable exception: no bootstrap. The reason for this should be obvious if we turn back to Fig. 9-15. In this figure we maintain low $R_{DS(on)}$ by bootstrapping the gate from the unity-gain amplifier, which, in turn, is sampling the analog signal voltage. Quite obviously, we do not wish to maintain a low $R_{DS(on)}$ if we are using the power MOSFETs as a VCR. $R_{DS(on)}$ control must remain under our control, *not* under the control of the analog signal. We need the in-phase analog signal to modulate the gate but not to take away control.

9.13 BENEFITS AND PROBLEMS IN USING POWER FETS AS ANALOG SWITCHES

Power FETs are the best choice for analog switches from among all other solid-state devices for a number of reasons that we have touched upon in this chapter. Bipolar transistors require substantial base current drive, and, like the FET, being a three-terminal device this base current finds its way into the signal path and results in what is known as "offset," which is simply an error signal developed by the base–emitter current flowing through the load.

The input characteristics of the power MOSFET offer us not only a prime advantage but also a major disadvantage, both of which we have described in this chapter. The advantage, of course, is that being a voltage-controlled switch we have no offset and we do not load the drivers. The latter makes possible direct drive from many forms of logic and allows us to fan out from one driver to an unlimited number of power MOSFET analog switches. The disadvantage we discovered is the limited gate-to-source breakdown voltage, which inhibits us from switching high-level analog voltages.

The disadvantage of the body–drain diode forces us to pair up power MOSFETs for analog switching, and with the newer ultralow-ON-resistance MOSFETs, although they present a disadvantage cost-wise, we have an unbeatable combination.

REFERENCES

Analog Switches and Their Application. Santa Clara, Calif.: Siliconix incorporated, 1976.

DOYLE, JOHN M., *Digital, Switching and Timing Circuits.* North Scituate, Mass.: Duxbury Press, 1976.

ten

Using the Power FET in High-Frequency Applications

10.1 INTRODUCTION

We may remember from Chap. 1 that both the SIT and the short-channel MOSFET were developed to solve the "high-frequency problem," so it would seem quite proper then that once having solved this problem, we would see them used in high-frequency applications.

For a few years during the mid-1960s we saw scattered evidence of Teszner and Zuleeg's analog transistor, but when Signetics introduced the short-channel, small-signal, high-frequency planar DMOS in 1969, the analog transistor all but totally dissappeared. This small-signal DMOS transistor struggled for nearly a decade but it, too, failed to gain wide acceptance in industry as a high-frequency transistor. Its demise, however, was not a failure, for it supplied the technology upon which was built the power VDMOS that we have today.

The first V-groove vertical short-channel power MOSFET (VMOS) used for high-frequency applications, the VMP4, was introduced by Siliconix incorporated in 1976 and remained unchallenged for four years.

An assembly of some of these early high-frequency power FETs is shown and identified in Fig. 10-1, together with some of those that are in wide use today.

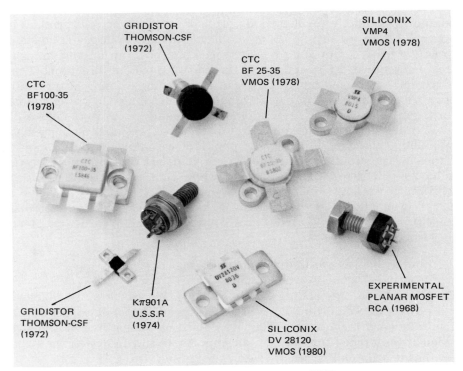

Figure 10-1 High-frequency power FETs.

The high-frequency power MOSFET market is quite different from any other. In the previous five chapters we focused our attention on several applications in which power FETs could find useful service. We must, however, realize that for most of these varied applications, a major redesign effort would generally be required, since we would be replacing either thyristors or power bipolar transistors, which operate on quite different principles. This is not always the situation in the high-frequency arena. Although the power MOSFETs do, indeed, operate on different principles, at high frequencies we find that retrofit is far less tedious and that the benefits are often worth the trouble. Market direction appears to be toward broadband systems, with the military tilting to spread-spectrum communications while the radio common carrier (RCC) market, still heavily committed to FM, is beginning to think sideband (SSB) because of the possible need for quasi-synchronous communications now being studied for many major metropolitan areas.

When Teszner, Zuleeg, and other early investigators were developing a high-frequency J-FET technology, their goals were primarily to improve on the gain–bandwidth performance, develop a high-speed component suitable for inclusion within an integrated circuit, and, as it appears today, to satisfy an insatiable desire to develop a solid-state "vacuum tube." Our goals have changed and now our interest in high-frequency power FETs appears to lead in two different directions. One is to seek higher and higher power at frequencies entering into the microwave region, for example, to develop an all solid-state power source for the microwave oven. The other direction aims to fulfill the needs of the communications market—a transistor of exceptional stability and reliability, allowing broadband operation with low baseband noise. Those investigators pursuing the former goals are developing the static induction transistor (SIT), and those whose goal is the latter have sided with the power MOSFET, the VMOS and the DMOS.

Power FETs have several distinct advantages over their bipolar transistor counterparts besides those that we studied in earlier chapters.

10.2 ADVANTAGES OF THE HIGH-FREQUENCY POWER FET

There are three outstanding features shared among both the SIT and the power MOSFET that have provided the enthusiasm to develop a high-frequency technology. First we itemize these features—not in any order of preference for they are all important—and then we review the significant benefits that we can derive from each one.

1. Power FETs do not appear to suffer from the characteristic voltage-frequency syndrome to the extent that we observe for the typical VHF bipolar transistor. That is, unlike the bipolar transistor, we do not observe a progressive decrease in breakdown voltage as we design for higher-frequency performance.
2. Another feature that we found to have contributed to the popularity of the power FET is its thermal characteristics. This has allowed us to parallel an infinite number of cells without the need of ballasting resistors.
3. Power FETs, as bulk semiconductors, lack both the bipolar transistor's base–emitter and base–collector diodes, which for the bipolar transistor are sources of noise.

The significance of some of these features may not be readily apparent to us without further discussion.

According to E. O. Johnson (see the References concluding this chapter), a bipolar transistor's voltage-frequency performance is set by the product of its breakdown voltage and its minority-carrier saturated drift velocity. In perhaps simpler language we are told that our upper frequency limit f_t depends on the transit time of the electrons traversing from emitter to collector. This frequency can be derived from the equation

$$f_t = \frac{l}{2\pi\tau} \tag{10.1}$$

where τ is the average time for a charge carrier moving at an average velocity v, to traverse the emitter-to-collector distance l. We should take a moment to compare and note the similarity between Eq. (10.1) and Eq. (2.7). In Chap. 2 we anticipated that our maximum operating frequency would be dependent upon the channel length and that breakdown voltage, drain current, transconductance, and channel resistance would all be affected. We anticipated the limitations posed by Eq. (10.1), but *we identified a partial solution*: use the short-channel offered by both DMOS and VMOS technology. At very high frequencies, beyond UHF, we find that the transit time within the drain-drift (epi) region must be considered. As we reduce the epi thickness to shorten the transit time, our breakdown voltage drops.

All high-frequency power bipolar transistors are designed by combining multiple cells to achieve the desired power level. As a result of a variety of defects stemming from both the material and the processing and because of the *regenerate* thermal properties of the bipolar transistor, some emitters will tend to draw a disproportionate amount of current, which could easily lead to thermal runaway and catastrophic destruction. The cure universally taken by manufacturers is to place ballast resistors in series with each emitter. If the emitter current begins to increase, we have an increased voltage drop across the ballast resistor, which forces a drastic reduction in this current, thus stabilizing the transistor. Power FETs are also built from many cells. However, since all FETs have a *degenerate* thermal characteristic, it matters very little how irregular each source diffusion is.

We saw in Chap. 1, especially from Fig. 1-5, that a transistor consists essentially of two diodes with a common base. The emitter–base diode is biased in the forward direction so that the emitter injects minority carriers into the base, whereas the base–collector is reverse-biased so that the collector extracts minority carriers from the base. Diodes are notorious sources of noise, and although we have several sources of noise in a bipolar transistor, by far the predominant noise generators are the emitter shot noise diode and the collector noise diode. The study of noise is far too unwieldy and complex for our present study, and the reader is invited to research the literature offered in the References concluding this chapter. We shall offer a few salient points with regard to noise in bipolar transistors before moving on to modeling the RF power FET.

There are, to be sure, many sources and forms of noise that occur in bipolar transistors as well as, of course, in power FETs. We can classify these forms of noise as follows:

1. Generator-recombination noise, which, as the name implies, is caused by spontaneous fluctuations within the semiconductor; a principal source of noise in MOSFETs, and a major source of noise in diodes which is often associated with *shot noise*.

2. Diffusion noise results, in part, simply because of inconsistencies of the diffusion process: as the temperature fluctuates, we find what is recognized as *thermal noise*. It can also contribute to shot noise in diodes.

3. Modulation noise may conceivably be a major source of noise in high-frequency power transistors since it is partially caused by current fluctuations resulting from a modulation mechanism, such as a predriver carrying some type of intelligence, such as audio modulation.

If we were to compare the broadband noise spectra of a bipolar transistor to that of an equivalently rated power FET, the apparent improvement in baseband noise would be obvious as shown in Fig. 10-2.

Power MOSFETs suffer principally from what is recognized as generator recombination noise and in particular from the effect of fast surface states which are located at the silicon–silicon dioxide interface. These surface states are very sensitive to voltage variations, as well as

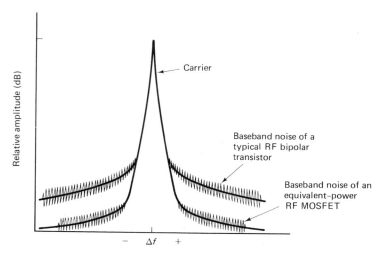

Figure 10-2 Comparing the broadband baseband noise spectra of the RF power bipolar transistor to that of a short-channel RF power MOSFET shows a marked improvement of greater than 20 dB.

being extremely sensitive to the crystallographic plane. Since <111>-orientation silicon exhibits the fastest surface states and <100>-orientation silicon the slowest, we can deduce that VMOS power MOSFETs should exhibit higher noise than a DMOS power MOSFET. This noise would, in all probability, be what is recognized as *flicker* or *1/f noise*. Should we choose to investigate the extensive literature of noise we would become aware of the distinction made between thermal (or Johnson) noise, flicker noise, and shot noise.

10.3 THE HIGH-FREQUENCY MODEL

In Chap. 4 we discussed modeling extensively but did not dwell on the high-frequency model, which we did show in Fig. 4-34 and have reproduced in Fig. 10-3. Generally, it is desirable to use the simplest model commensurate with the problem. In switching and low-frequency applications, simple models can be very effective for most circuit designers. Unfortunately, at high frequencies a simple model is often grossly inadequate in describing performance, as we saw in Fig. 4-35, and even complex models can fall short in providing accurate data. The design engineer who uses models will find that, at best, they offer the opportunity of near misses rather than wild shots, and generally some trimming of the circuits is necessary. Perhaps the major shortcoming for most bipolar transistor models is that we model small-signal parameters even when our need is for a power device. Fortunately, a benefit of the short-channel power MOSFET finds its small-signal parameters closely matched to its power parameters. This rather fortuitous benefit arises from the voltage dependency of the parasitic capacitances as well as from the absence of minority carriers. Modeling is simplified if we can comfortably consider that all the parasitic elements, including the capacitances, are constants, which, for the short-channel power MOSFET, is true provided that we maintain fixed voltages.

The schematic model shown in Fig. 10-3 represents both VMOS and VDMOS technologies and, although reasonably "exact," does suffer at low frequencies. At frequencies above 80 MHz, the model *is* exact. As we saw in Chap. 4, the principal omission found in most MOSFET models is the parasitic bipolar transistor straddling the body diffusion, with the collector tied to the drain and the emitter to the source, the base and body being one and the same. Early in the evolution of the short-channel power MOSFET, we saw repeated instances of defective designs, where this parasitic *npn* (for an *n*-channel MOSFET) bipolar transistor could be activated. Today's designs have all but totally eliminated any possibility of activation. Nonetheless, the parasitic bipolar transistor exists, not as an active transistor but as a static, or nonactive, transistor.

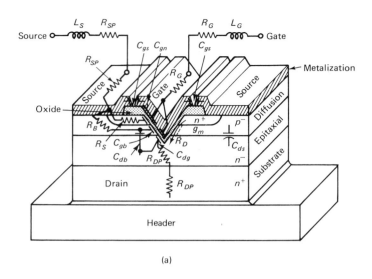

(a)

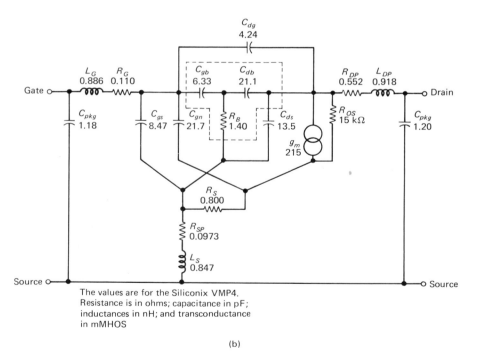

The values are for the Siliconix VMP4.
Resistance is in ohms; capacitance in pF;
inductances in nH; and transconductance
in mMHOS

(b)

Figure 10-3 High-frequency model of a short-channel power MOSFET. (a) The physical model of VMOS. (b) The electrical model showing values for the Siliconix VMP4. (Courtesy of *RF DESIGN* © 1979 Cardiff Publishers.)

This schematic model (Fig. 10-3) was drawn by carefully following the signal path through the physical model, shown earlier in Figs. 2-28 (VDMOS) and 2-29 (VMOS). C_{gs} differs from C_{gn} in that the former is the field capacitance, whereas the latter is the parasitic capacitance between the gate metal and the n^+ source diffusion. The static *npn* bipolar transistor contributes a feedback mechanism consisting of C_{gb}, C_{db}, and R_B. Only two parameters could not be modeled directly from the physical model; these were g_m, the *transconductance*, and R_{OS}, the reciprocal of the *output conductance*.

Using this schematic model, we can assign values of each parameter with a minimum of effort by using nodal analysis available from a circuit optimization program called COMPACT.[1] The nodal analysis interconnection network is shown in Fig. 10-4, and the complete program to solve the parameter values is offered in Fig. 10-5. Since the details of this program are available from several computer service firms, we will not spend time here analyzing its operation. We should point out, however, the need for accurate estimation of parameter values. You must know "where you're coming from" or else be prepared to spend a considerable amount of money on computer time. Fourteen variables were inputted into the program, which are identified in Fig. 10-5 as negative amounts. All the other terms were either measured, such as the *S*-parameters, or had a high probability of being sufficiently accurate to preclude possible error. The results of this modeling program are shown for the Siliconix VMP4 VMOSFET in Fig. 10-6, where we see surprisingly good correlation between measured and calculated parameters. We must remember that these parameters are small-signal parameters, and although the model is certainly not a simple one, if we choose to substitute a less complicated one, we *must* include the elements of the parasitic *npn* bipolar transistor if we expect to model gain parameters. A simple model, valid for operation below 60 MHz, is shown in Fig. 10-7.

10.4 PARAMETERS AFFECTING THE PERFORMANCE OF POWER MOSFETS

Each year we see the steady advance of increasing power and increasing frequency as the technology learns to overcome past problems. We are aware that the range of frequencies over which a transistor, be it a bipolar transistor or a power MOSFET, performs is limited by certain inherent parameters. In this section we identify some of the principal parameters and also attempt to show how they affect performance and why.

[1] COMPACT is the registered trademark of Compact Engineering, Palo Alto, Calif.

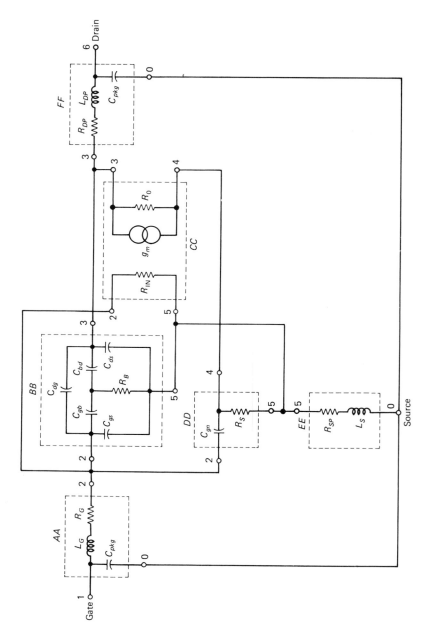

Figure 10-4 Nodal analysis for COMPACT network synthesis. (Courtesy of *RF DESIGN* © 1979 Cardiff Publishers.)

```
CAP  AA  PA  1.18                        CON  BB  T3  2.00          3.00  5.00
SRL  BB  SE  -0.110        -0.886        CON  CC  T4  2.00          5.00  4.00  3.00
CAX  AA  BB                              CON  DD  T3  2.00          4.00  5.00
CAP  BB  PA  -8.47                       CON  EE  T2  5.00          0.00
CAP  CC  SE  -6.33                       CON  FF  T3  3.00          6.00  0.0
RES  DD  PA  -1.40                       DEF  AA  T2  1.00          6.00
CAP  EE  SE  -21.10                      TWO  BB  S1  50.0
CAP  FF  PA  -13.50                      SET  AA  BB
CAX  BB  FF                              PRI  AA  S1  50.0
CAP  CC  SE  -4.24                       END
PAR  BB  CC                              200
GEN  CC  VC  0.100E + 11  0.150E + 05  215   END  0.77  -146.3  2.14  57.5  0.035  -6.0  0.759  -146.6
CAP  DD  SE  -21.7                       END
RES  EE  PA  -0.800                      END
CAX  DD  EE                              0.001
SRL  EE  SE  -0.973E - 01  -0.847        1  1  1  0.627
SRL  FF  SE  -0.552        -0.918        END
CAP  GG  PA  1.20
CAX  FF  GG
CON  AA  T3  1.00          2.00 0.0      EOF:
```

Figure 10-5 Elemental values for the Siliconix VMP4 for use in COMPACT analysis (Fig. 10-4). (Courtesy of *RF DESIGN* © 1979 Cardiff Publishers.)

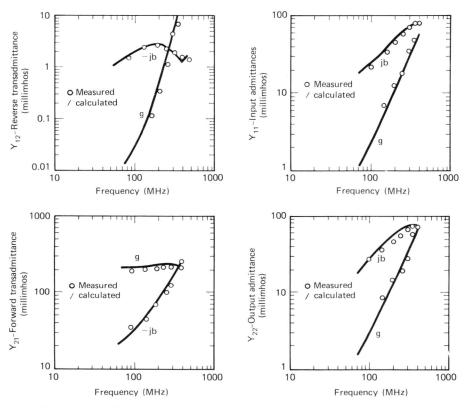

Figure 10-6 The small-signal modeled y-parameters of the Siliconix VMP4 compared with measured data show close agreement above 80 MHz.

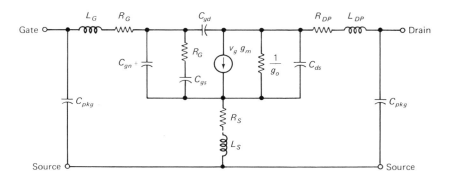

Figure 10-7 Simple power MOSFET model (excluding the parasitic *npn* bipolar transistor elements).

One of the most important factors in the power-handling ability of a power bipolar transistor is the concentration of injected current about the periphery of the emitter. For this we can derive what we might call a *figure of merit*,

$$\text{F.M.} = \frac{\text{emitter periphery}}{\text{base area}} \tag{10.2}$$

Unlike the definition offered in Chap. 2 [Eq. (2.5)], here our figure of merit identifies the ratio of *power output* (emitter periphery) to *upper frequency of operation* (base area).

We have not backed ourselves into a paradoxical corner by defining figure of merit two ways, that is, for a power bipolar transistor as the ratio of power output to upper frequency of operation, as shown in Eq. (10.2), and then, as we saw in Eq. (2.5), the ratio of forward transconductance to input capacity. If we were to rewrite Eq. (10.2) expressly for a power MOSFET, we would have

$$\text{F.M.} = \frac{\text{gate periphery}}{\text{drain area}} \tag{10.3}$$

Gate periphery means a wider gate or, in other words, a wider channel, W. For the power bipolar transistor an extended emitter periphery means higher power, whereas for the power MOSFET it means higher transconductance. It would be like paralleling power FETs. Similarly, if we reduce the drain area, we also reduce the output capacity, consisting of both C_{ds} and C_{dg}. The latter term is important, as C_{dg} contributes directly to C_{in} as a result of Miller effect. Consequently, we can "rewrite" Eq. (10.3), and the similarity to Eq. (2.5) is *not* coincidental.

$$\text{F.M.} \simeq \frac{g_m}{C_{in}} \tag{10.4}$$

This figure of merit that we have used here as well as in Chap. 2 is referred to as the *gain–bandwidth product* f_t, defined in Eq. (10.1) as being inversely proportional to transit time. If we now align the three equations that we have identified for f_t, we might be able to recognize more clearly how certain parameters relate.

$$f_t = \frac{g_m}{2\pi C_{in}} = \frac{v_s}{2\pi L} = \frac{1}{2\pi\tau} \tag{10.5}$$
$$\text{(a)} \qquad \text{(b)} \qquad \text{(c)}$$

Equation (10.5a) identifies the importance of high transconductance g_m, and low input capacity C_{in}; Eq. (10.5b) equates f_t with high carrier velocity through a short channel, L; and Eq. (10.5c) repeats what we learned earlier in this chapter with regard to transit time.

In Chap. 4 [Eq. (4.11)] we saw that the ratio of channel width W to channel length L determined the forward transconductance g_m:

$$g_m = \frac{W}{L} \qquad (10.6)$$

If we now stop to consider the impact of these equations, we will be able to draw some conclusions regarding what parameters affect performance and why.

In Eq. (10.3) we see a small drain area contributing to improving the f_t. What we are really seeing is a reduction in the p-channel, $n-$ epi area, which, in turn, reduces the output capacity. But output capacity is, as we saw earlier, a combination of C_{ds} plus C_{dg}; therefore, any lowering of C_{dg} helps to lower the effective input capacity C_{in} (by virtue of the reduced Miller effect).

The three parts of Eq. (10.5) give us insight into the relationship between channel length L and its effect on C_{in}. The shorter the length, the lower the source-to-gate capacitance. Furthermore, as the channel length shortens, carrier velocity may approach or become saturated and we see a reduction in transit time τ.

It should be clear that we have two ways to increase the frequency of a power MOSFET: to reduce C_{in} or raise g_m. Before we turn our attention toward those parameters involved in raising the power, let us conclude with a final equation that defines the *maximum stable gain* (MSG) of a power FET:

$$\text{MSG} = \frac{g_m + j\omega C_{dg}}{j\omega C_{dg}} \qquad (10.7)$$

Again, we see the dependence of performance tied directly to C_{dg}.

Ignoring for the moment the frequency response of the power FET, we can intuitively conclude that to improve the power handling of the FET, we have got to reduce the intrinsic losses: electrical resistance $R_{DS(on)}$ and *thermal* resistance R_Θ. We have got three options: (1) reduce the epi thickness (drain-drift region), (2) increase channel width W, and (3), lower the thermal resistance by either increasing the area or thinning the wafer. The first option brings with it two major drawbacks. A thin epi lowers the ruggedness of the device, and if we push for higher-frequency operation, we find that our FET's operation behaves similar to the Johnson effect; that is, as frequency increases, the breakdown voltage decreases. Our second option is nothing more than paralleling power FETs. As we add more active devices, the transconductance g_m increases proportionally with C_{in}. Theoretically, we see no deleterious effects on operating frequency, but in truth, we do have some degradation.

The most viable option we have to reduce the thermal resistance is to design our power transistor using many small cells rather than a few large cells and to spread them out sufficiently to provide thermal equilibrium across the entire chip. In the final manufacturing phase before scribe (or saw) we can backlap the wafer to reduce the overall thickness possibly by as much as 50%. It is not unusual for high-frequency power bipolar transistors to be only 6 mils thick.

There are many parameters that affect the performance of power MOSFETs at high frequency that we did not describe in this section. What we tried to do here was not to narrow our perspective to a single application but rather to take a "broad brush" at what parameters affect high-frequency performance in general. We did not discuss which geometry was superior, for example, DMOS or VMOS, nor did we identify any particular topology. These decisions we must leave with the vendor. As we proceed through the chapter, we introduce specific applications and parameters that either assist us in achieving our goal or detract us from complete success.

10.5 SYSTEMATIZING RF POWER DESIGN

Irrespective of our application, there are several fundamental factors that we must consider before we can arrive at a successful design. We need to identify the *class of operation*; are we trying for high linearity single-sideband amplification, or is our goal high-efficiency power amplification? As with any power amplifier, the class of operation has an important bearing on the power output, linearity, and operating efficiency. We need to establish a *load line* based on the desired output power, the source voltage, and the transistor we wish to use (the transistor itself does not enter the calculation per se, but we have got to be sure that it will handle the power). We have to establish the *stability* of our amplifier, which may, in turn, affect how we go about matching the input and output ports and whether or not we need to neutralize the transistor. In many ways it matters very little whether we plan to use power MOSFETs or bipolar transistors as we begin our design. That is partially the reason why it is often easier for us to retrofit a MOSFET into an existing high-frequency design than it might be to retrofit a MOSFET into a switching power supply, especially into an existing motor controller. We should note, however, that any retrofit is unilateral; that is, we can generally always replace a power bipolar high-frequency transistor with a power MOSFET, but we may not be able to reverse the switch—we will find out why shortly.

Power MOSFETs operating at low frequencies, for example in video (10 kHz through 10 MHz) and in the HF band (3 through 30 MHz), exhibit what amounts to a nearly pure capacitive input impedance. The shunt resistive component is extremely high, since we are "looking" into an MOS gate dielectrically isolated. If we accept as a general rule that high-frequency power MOSFETs have nearly infinite input resistance at low frequencies, we can immediately identify a problem if we would like to design an amplifier with a good input match. A pure reactance does not absorb power; it merely reflects it. To absorb power we must have a resistive load, and furthermore, to develop a voltage across the MOSFET's gate-to-source, we need a shunt resistive element. Because our short-channel power MOSFET offers a reasonably constant saturated transconductance irrespective of drain current and frequency (within reasonable limits), we can begin with the general expression for power gain for an active two-port network and derive a novel equation that will satisfy our need for a gate input resistance.

We begin with the general equation

$$G_P = \frac{|y_{21}|^2 \ \text{Re}(Y_L)}{|Y_L + y_{22}|^2 \ \text{Re}\left(y_{11} - \frac{y_{12}y_{21}}{y_{22} + Y_L}\right)} \tag{10.8}$$

To further simplify our derivation for low-frequency applications, we can disregard the reactive elements, and since our major feedback mechanism is the now-rejected capacitive reactance, we can set y_{12} equal to zero. Equation (10.8) reduces to

$$G_P = \frac{(g_m)^2 \ (1/R_L)}{[(1/R_L) + g_{22}]^2 \ g_{11}} \tag{10.9}$$

The term g_{22} for a saturated power MOSFET is very low, certainly in the millimhos; consequently, its reciprocal, R_{22}, would be quite a high value, certainly higher than R_L, the load resistor. The term g_{11} is really nothing more than $1/R_S$, the reciprocal of the gate input resistance. Reducing Eq. (10.9), we have

$$G_P = (g_m)^2 \ R_S R_L \tag{10.10}$$

where G_P is a numerical value which can be further reduced to a more convenient form,

$$G_P \ (\text{dB}) = 10 \log (g_m)^2 \ R_S R_L \tag{10.11}$$

Now, we can easily solve for R_S:

$$R_S = \frac{10^{\text{dB}/10}}{g_m{}^2 \ R_L} \tag{10.12}$$

The load resistance R_L may be calculated using a well-known large-signal equation,

$$R_L = \frac{(V_{DD} - V_{DS(\text{sat})})^2}{2P_{\text{out}}} \tag{10.13}$$

We must not lose sight of the restrictions imposed upon our use of Eq. (10.12). Should the reactive elements of our transistor become significant, this equation will lose its effectiveness and we will have to return to Eq. (10.8). Equation (10.13) is good, however, irrespective of the reactive elements, but our load impedance would have to include the conjugate of the parallel equivalent output capacitance and R_L.

If our power MOSFET stage is one-half of a push-pull arrangement, we must be aware of one major difference that sets the FET apart from a comparable push-pull bipolar transistor amplifier. A negative-going voltage waveform feeding the base of a common-emitter bipolar transistor draws little current, simply because the base-emitter junction appears as a back-biased diode. This is quite obviously not so for the FET. If the power FET's input impedance has a resistive (and/or a low reactive) component, *current will flow*. Although this may not appear particularly profound, it does force us to reexamine the input transformer primary-to-secondary ratio.

Another interesting aspect that we should note with power FETs is the ease of biasing. To establish a quiescent drain current, we do not need a bias current as we would require for a bipolar transistor. Consequently, decoupling our bias supply becomes quite easy, especially if we use a large-value carbon-composition resistor.

Earlier in our discussion of the high-frequency model (Sec. 10.3), we laid claim to the stability of the small-signal parameters, even at large signal levels. This is especially true for Class A designs and it holds reasonably well for Classes B and C insofar as the input impedance is concerned. Generally, the output load line is derived from Eq. (10.13), but the transistor's output capacity can be closely approximated from the small-signal Y-parameter, y_{22}.

In push-pull arrangements we must, as we would for push-pull power bipolar transistors, take special care in the design of the output transformer stage, especially if we seek wideband performance using transmission-line transformers. These are very effective for impedance matching over wide bandwidths, but their available impedance ratios are constrained to be squares of integer ratios [e.g., $(\frac{1}{2})^2$, $(\frac{1}{3})^2$, $(\frac{1}{4})^2$, $(\frac{2}{3})^2$, etc.]. If we believe that we need a more exact match than what can be achieved with these transformers, we can resort to lumped LC impedance transforming networks.

The technique of proper matching of push-pull amplifiers has been a problem shrouded by doubt and suspicion. There appears to be at least two interrelated problems. The first problem relates to achieving a proper impedance match between the transistors and the transformer; the second problem concerns the efficient transfer of power from the transistors to the load. The problem manifests itself by the class of service. A Class A push-pull amplifier operates with both transistors active for the full cycle (360°), whereas any other class of service has neither transistor active for the full cycle. For example, the transistors in a Class B stage operate at a 50% duty cycle and for Class C, considerably less. If we were to consider that our transformer is a three-port device having two ports for inputting energy from the transistors and one port for discharging energy into our load, the problem becomes visible when we recognize that for Class A service we have two active generators operating simultaneously, whereas for any other class we have only one. We need not enter into further discussion to appreciate that a transformer designed for dual input does not necessarily work at peak performance with only one active input. Theoretically, at least, if we design an optimized Class A push-pull stage, we cannot simply change the bias to perform in another class. Figure 10-8 offers typical output transformer combinations suitable for Class A service and Class B. We should take special pains to understand the current flow diagrams to appreciate the problem.

Our desired drain load impedance for a Class A push-pull amplifier is derived from Eq. (10.13). Since both transistors operate continu-

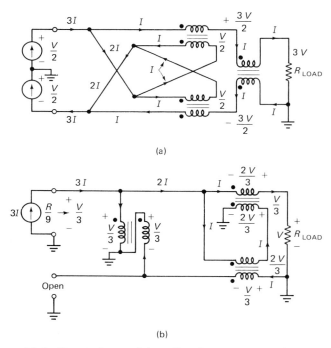

(a)

(b)

Figure 10-8 Comparison of broadband output transformers suitable for Class A (a) and Class B or C (b). [(b) From *Solid State Radio Engineering* by Krauss, Bostian, and Raab, © 1980. Courtesy of John Wiley & Sons, Inc., New York.]

ously, we must calculate a *drain-to-drain* load impedance using the equation

$$R'_L = \frac{2(V_{DD} - V_{DS(\text{sat})})^2}{P_{\text{out}}} \tag{10.14}$$

Higher-class push-pull amplifiers (such as Class B and C) can, of course, use Eq. (10.13) to establish R_L. We might note that R'_L of Eq. (10.14) is quadruple the R_L of Eq. (10.13).

An important aspect of amplifier design must concern *stability*: the ability of the amplifier to be free of unwanted spurious oscillations. We must be able to design power amplifiers not only for optimum gain and output power (and efficiency), but we must be sure that we have an *amplifier* when our design is completed and not an *oscillator*. It takes more than shielding to assure ourselves of this.

Without our becoming involved in detail quite beyond the intention of this book, we should be aware of two equations that define stability. The first defines the potential stability of the transistor, which, of course, can either attract or distract the designer from making a selection. This equation is the Linvill stability factor, which is a measure of the transistor's stability when both input and output ports are left open-circuited.

$$C = \frac{y_{21}y_{12}}{2g_{11}g_{22} - \text{Re}(y_{21}y_{12})} \tag{10.15}$$

If C is less than 1, the device is unconditionally stable, but if we calculate C greater than 1, the transistor is potentially unstable and our circuit might oscillate.

The second equation relative to stability focuses our attention on the amplifier circuit itself with the intention of defining the source and load admittances to assure stability. This equation is the Stern stability factor, K:

$$K = \frac{2(g_{11} + G_S)(g_{22} + G_L)}{|y_{21}y_{12}| + \text{Re}(y_{21}y_{12})} \tag{10.16}$$

Here we see two new parameters, G_S and G_L, the former being the real part of the source admittance, the latter the real part of our load admittance. If K is less than 1, the circuit is potentially unstable.

An exhaustive study of the many facets of high-frequency design is quite beyond the overall scope of this book. What we have endeavored to accomplish in this section is to acquaint the reader with some of the salient features unique to power MOSFETs when we use them in high-frequency design. As we complete this chapter we will gain additional visibility and, hopefully, become convinced that the short-channel MOSFET will have increasing influence in the high-frequency market.

10.6 POWER MOSFETS IN LINEAR AMPLIFIERS

We probably should first disassociate ourselves from too narrow a perspective, believing that a linear amplifier means a linear *power* amplifier or a Class A or AB amplifier. True, we can believe all or part of the above, but there are options that we will touch on in this section. If our need is for small to medium power with high linearity, power MOSFETs excel. On the other hand, if we are trying to develop appreciable power and good efficiency—not a particularly easy task for a Class A or AB amplifier—for the purpose of, say, the amplification of *single-sideband* (SSB) signals, power MOSFETs used in the Kahn system work very well.

10.6.1 An Ultralinear Broadband Video Amplifier

In small-signal amplifier design we often find it convenient to replace the transistor by its equivalent model and use *computer aided design* (CAD) to establish a preliminary foundation for our circuit design. Although the exact model that we reviewed in Sec. 10.3 was rather complex for low frequencies, we found, much to our delight, that by using the Y-parameters derived from that model, our video amplifier design, shown in Fig. 10-9, worked the first time. To set both the input and output match and establish the gain, we used the following equations.

$$R_1 = \frac{(R_S R_L)^{1/2}}{2}\left((G)^{1/2} + \left\{G + 4\left[1 + (G)^{1/2}\,\frac{R_S + R_L}{2(R_S R_L)^{1/2}}\right]\right\}^{1/2}\right)$$

(10.17)

$$R_2 = \frac{R_S R_L}{R_1} - \frac{1}{g_m}$$

(10.18)

where R_S and R_L = source and load impedances, respectively

g_m = forward transconductance of the power MOSFET expressed in mhos ($\mathrm{Re}\,Y_{21}$ from the model)

G = numerical gain ratio desired for the stage, or $10^{\mathrm{dB}/10}$

Measurements on the completed video amplifier confirmed the results derived from CAD: an input VSWR less than 1.2:1 and a gain of 12 dB at 2 and 30 MHz. The bandwidth of the video amplifier extended

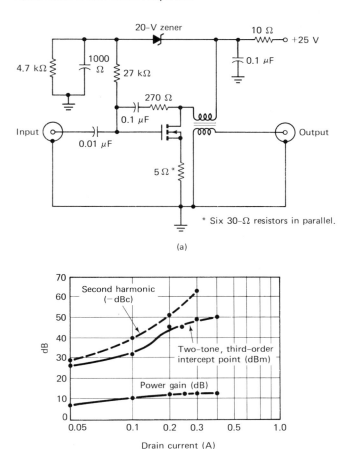

(a)

(b)

Figure 10-9 Ultralinear broadband Class A video amplifier (a) and performance (b). (Courtesy of *QST*, May 1979.)

from less than 50 kHz to 70 MHz. A two-tone, third-order intercept point followed the dynamic transfer characteristics of the MOSFET.

10.6.2 A 100-Watt Class AB Power Amplifier

The schematic shown in Fig. 10-10 provides a power gain of 12 dB with an output power of typically 100 W across the 30- to 88-MHz band. Since the input impedance of these Siliconix power VMOSFETs (DV2880) is very nearly a pure capacitive reactance at these frequencies, the resistors R1 and R2, tied from each gate to ground, are calculated from Eq. (10.12).   The broadband transformer, designed to

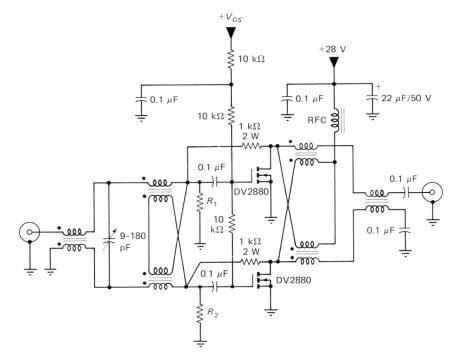

Figure 10-10 A 100-W broadband VMOSFET power amplifier using Siliconix DV2880 offers a power gain of 10 dB at a quiescent drain current of 2.2 A. (Courtesy of *RF DESIGN* © 1980, Cardiff Publishers.)

operate into a pair of identical loads simultaneously, offers a good match across the entire bandwidth.

Since the amplifier operates Class AB with a large conduction angle, we can use the simple 4:1 hybrid transformer and 1:1 balun arrangement to transform from a drain-to-drain load of 12.5 Ω to 50 Ω. Using the values R_S = 6.25 Ω, R_L' = 12.5 Ω (R_L = 3.12 Ω), g_m = 0.7 mho, and from the vendor's data sheet, the worst-case value for g_{22} of 0.026 mho, we are able to calculate the gain using Eq. (10.11).

Potential instability resulting from excessively high output mismatch is completely cured by the addition of 1-kΩ feedback resistors. Before moving on we should stop and consider the problems of stability. Since our amplifier has finite source and load impedances, we can calculate circuit stability using the Stern stability factor equation (10.16). Since our broadband amplifier operates from 30 to 88 MHz, for our illustration we can take the center frequency of 50 MHz as our basis for computation. From the DV2880 data sheet we itemize the Y-parameters at 50 MHz.

$$Y_{11} = 4.4 + j50.3 \qquad Y_{12} = 0.4 - j5.6$$

$$\text{(all in mmhos)}$$

$$Y_{22} = 18.3 + j40.7 \qquad Y_{21} = 700 - j81.5$$

and

$$G_S = \frac{1}{6.25} \text{ mho} \qquad G_L = \frac{1}{12.5} \text{ mho}$$

Substituting into Eq. (10.16), we find, as we fully expected, that the Stern stability factor K exceeded 1, identifying complete circuit stability.

10.6.3 A Linear and Highly Efficient Power Amplifier

Those of us familiar with the various classes of amplifier service will admit that a linear Class A amplifier is not known for its efficiency, whereas a Class C, or higher class, has excellent efficiency but generally very poor linearity. A further problem of conventional Class A or AB power amplifiers is that they will amplify lower-level spurious frequencies and noise, so it is to our advantage to resolve linear amplification using other than Class A or AB amplifiers. Conventional high-level modulation techniques have three fundamental shortcomings that restrict their use: (1) expensive to use because of the cost of the audio modulator; (2) the double-sideband output signal requires excessive bandwidth; and (3) inefficient utilization of the total available dc power.

A novel approach using power MOSFETs has been proposed by two British research scientists using a concept developed by Kahn (see the References concluding this chapter) that allows us high efficiency and excellent linearity. Their circuit, using the Siliconix VMP4, is shown in Fig. 10-11, and we can see the remarkable improvement in linearity as a result of this envelope feedback scheme by examining the graph in Fig. 10-12. What is especially spectacular is the two-tone intermodulation performance shown in Fig. 10-13.

10.7 HIGHER EFFICIENCY FROM SHORT-CHANNEL MOSFETS

Although a short-channel MOSFET has finite and performance-limiting parameters such as $R_{DS(on)}$, parasitic capacitances, and a lower transconductance than an equivalent power bipolar transistor, we have focused our attention more than once on the fact that a FET is a bulk semiconductor and exhibits no minority-carrier storage time. This unique feature of the power FET makes it a candidate for high-

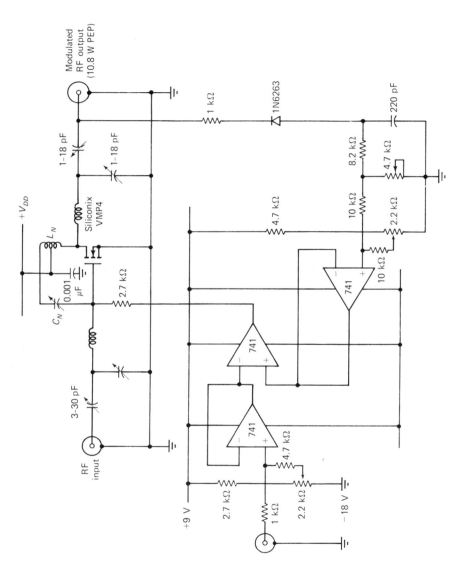

Figure 10-11 High-efficiency AM transmitter, as developed at Bath University, using short-channel MOSFETs. [Courtesy of IEE Conference Publication 162 (1978) UK.]

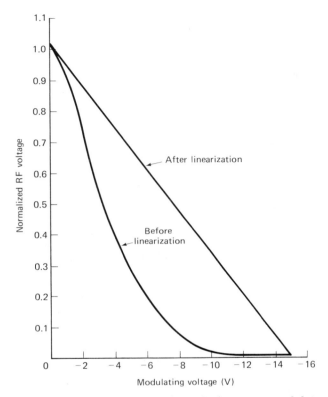

Figure 10-12 Transfer characteristics of the gate modulator (Fig. 10-11), showing the remarkable improvement resulting from envelope feedback. [Courtesy of IEE Conference Publication 162 (1978) UK.]

efficiency *switching* power amplifiers. As we read in Chap. 5, we find that little power is dissipated during the switching time if we can forget storage time. *And we can for a power FET.* Of course, a power FET is not perfect; it does have an inherent resistance, as we saw in Eq. (4.14).

Without fear of oversimplification, we can perhaps reach an understanding of how a zero-storage-time power FET can improve efficiency through switching. A Class A power amplifier, operating from a fixed supply, draws drain current for the entire 360° of the ac drive and maximum efficiency cannot exceed 50%. A Class B power amplifier draws drain current for only 180° of the ac drive. Since efficiency relates to the power consumed by the amplifier, if our amplifier now draws current half time, our efficiency should improve proportionally. Theoretically, we can attain an efficiency of 78.5%. Now a Class C RF power amplifier draws current for even less than 180°; consequently, its efficiency can climb even higher, theoretically at least to

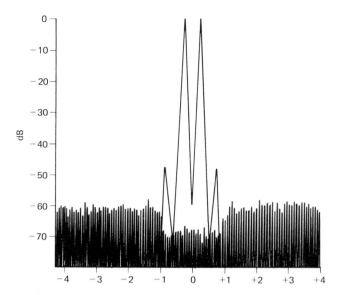

Figure 10-13 Graphical representation of two-tone, third-order intermodulation products of the high-efficiency AM transmitter with
envelope feedback.   [Courtesy of IEE Conference Publication 162
(1978) UK.]

100%. In a practical sense, however, a Class C amplifier peaks at about
85%. Now if we can control our circuit so that we are able to switch at
zero drain current, our efficiency should be 100%, and that is the
principle used in most high-efficiency RF power amplifiers. Our
advantage with power MOSFETs is in switching. If we can switch fast
enough—and saturate well enough—we can achieve very high efficiencies. When the transistor is nonconducting, no current flows and
no power is lost; when the transistor is hard ON and we have, theoretically, no voltage drop, we again have no power loss. Such a switching
amplifier is often identified as a Class D amplifier, and more often than
not, we can identify the class of operation by the presence of a push-
pull (complementary or quasi-complementary) transistors and a series-
resonant output tank circuit similar to that shown in Fig. 10–14. This
series-resonant tuned circuit offers low reactance to the fundamental
(switching) frequency and high reactance to all harmonics. The drain
current follows the response of our series-resonant tank and we find
a square wave of voltages built up at point *A*. Our fundamental, of
course, passes directly to the load, and since the harmonic currents are
negligible, we discover that we have, in effect, a reconstituted sine wave
of current appearing across our load and, according to Raab (see the

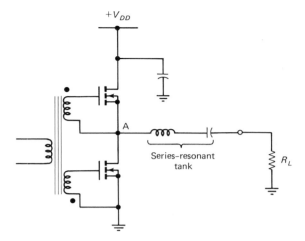

Figure 10-14 Output tank circuit of a Class D RF amplifier using totem-pole MOSFETs.

References), our output power is

$$P_{out} = \frac{2V_{DD}^2}{\pi^2 R} \qquad (10.19)$$

We should note that since our series-tuned resonant tank circuit must present a low reactance to the fundamental and high reactances to all harmonics, achieving broadband operation is not easy.   As we discovered in Chap. 5, our intrinsic "catch" diode becomes a valuable snubber to protect our transistors should we experience a reactive load rather than a resistive one.

   If in our investigations of various novel high-frequency power amplifiers we should stumble across a single output transistor coupled to a series-resonant tuned circuit, we may have before us the Class E power amplifier invented by the Sokals. The schematic of a 70-MHz 5-W Class E amplifier, shown in Fig. 10–15, offers better than 90% efficiency with a series-resonant tank circuit Q of 3.

10.8 POWER MOSFETS IN DISTRIBUTED AMPLIFIERS

The stage gain and bandwidth of a conventional amplifier is limited by the figure of merit, or gain–bandwidth product of the active element.   For any circuit we can increase the bandwidth at the sacrifice of gain, and vice versa.   To improve our gain–bandwidth product we have to do more than simply parallel transistors. Accepting whatever

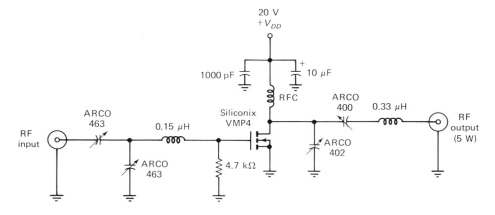

Figure 10-15 High-efficiency 70 MHz Class E RF amplifier. (A high-efficiency VMOS Class E amplifier. For construction of experimental Class E circuits, please obtain license under U.S. Patent 3,919,656 from N. Sokal, Design Automation, Inc., 809 Massachusetts Ave. Lexington, MA 02173.)

transistors are commercially available, we can achieve some slight improvement by clever circuit design, but for the most part our success is minimal. Equation (10.5a) identifies the classic problem. If we parallel active devices, both the transconductance and the capacity will rise together (actually, we will find the capacity rising faster because of strays and parasitics). We have to increase the transconductance and maintain a fixed capacity. Since most of us are not involved in transistor design, our only solution is the *distributed amplifier*.

A distributed amplifier overcomes the fundamental gain-bandwidth limitation by having the active elements straddle a pair of trans-

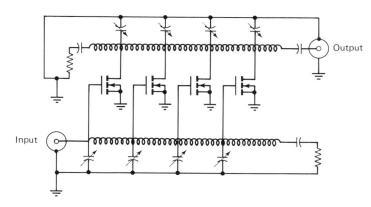

Figure 10-16 Basic distributed amplifier using *m*-derived low-pass filter elements to achieve bandwidths from dc to f_c (cutoff).

mission lines so that their transconductance add but their capacitances become an integral part of the transmission lines and, therefore, do not enter into the gain–bandwidth equation.   The design of distributed amplifiers basically hinges upon the configuration chosen for the input and output transmission (delay) lines.   A number of designs are possible, but the simplest is the image parameter, m-derived, low-pass filter shown in Fig. 10–16.   There are, however, a number of critical areas that need our careful attention.

For a distributed amplifier to operate sucessfully, the input and output transmission lines need not have equal characteristic impedance, but we must ensure that their phase velocities are equal.   The two equations of particular importance in the design of an m-derived low-pass distributed amplifier are:

$$\text{cutoff frequency:}\quad f_c = \frac{1}{\pi(LC)^{1/2}} \tag{10.20}$$

$$\text{phase constant:}\quad \beta = 2\pi f(LC)^{1/2} \tag{10.21}$$

Establishing the transmission line's characteristic impedance Z_0 depends to a great extent upon the input and output capacity of our power MOSFET.   The higher this capacity, the lower we must go in either Z_0 or f_c or both.   Since the characteristic impedance of our two transmission lines may be calculated from the equation

$$Z_0 = (\frac{L}{C})^{1/2} \tag{10.22}$$

we find that to establish equal phase velocity, we had best design our transmission lines by iteration using Eqs. (10.20) through (10.22).

TABLE 10-1   Series Inductance and Shunt Capacity Values

F_C	50 Ω	100 Ω	200 Ω	300 Ω	500 Ω	1000 Ω
50	127.3 pF	63.7 pF	31.8 pF	21.2 pF	21.2 pF	6.36 pF
	318.3 nH	637 nH	1.27 µH	1.9 µH	3.17 µH	6.36 µH
100	63.7 pF	31.8 pF	15.9 pF	10.6 pF	6.37 pF	3.18 pF
	159.3 nH	318 nH	636 nH	0.95 nH	1.59 nH	3.18 nH
200	31.8 pF	15.9 pF	7.95 pF	5.3 pF	3.18 pF	1.59 pF
	79.5 nH	159 nH	318 nH	0.48 µH	0.79 µH	1.59 µH
300	21.2 pF	10.6 pF	5.3 pF	3.5 pF	2.12 pF	1.06 pF
	53 nH	106 nH	212 nH	0.315 µH	0.53 µH	1.06 µH
500	12.7 pF	6.36 pF	3.18 pF	2.12 pF	1.27 pF	0.64 pF
	31.8 nH	63.6 nH	127 nH	0.19 µH	0.318 µH	0.64 µH
1000	6.36 pF	3.18 pF	1.59 pF	1.06 pF	0.64 pF	0.32 pF
	15.9 nH	31.8 nH	63.6 nH	95.4 nH	0.16 µH	0.32 µH

This table provides, for any cutoff frequency (f_c) between 50 MHz and 1 GHz, elemental values of series inductance and shunt capacity that will insure equal phase velocity for any characteristic impedance from 50 to 1000 Ω.

This we have done in Table 10-1 for the low-pass distributed amplifier of Fig. 10-16.

We can calculate the ideal voltage gain quite easily if we first recognize that the signal current in the drain line travels in two directions: toward the load as well as back toward the terminating resistor. Consequently, our voltage gain is, ideally,

$$A_V = \tfrac{1}{2} n g_m (Z_g Z_d)^{1/2} \qquad (10.23)$$

if we presume that we have no dissipative losses.

In a distributed amplifier our principal dissipative loss results from the gate conductance, Re Y_{11}. Since we have effectively bypassed the conventional figure of merit, or gain–bandwidth of our power MOS-FET, we should perhaps consider a more appropriate figure of merit for distributed amplifiers. If we were well versed in the design and fabrication of distributed amplifiers, we would know that the major short-coming that limits performance is the input conductance of the active elements; we cannot continue to increase our gain by indiscriminate adding of active elements along the transmission line. If we choose a nominal stage gain of approximately 8 dB, our cutoff frequency f_c [see Eq. (10.20)] can be calculated as

$$f_c = 0.6(f) \frac{g_m}{\text{Re } Y_{11}} \qquad (10.24)$$

where (f) is the frequency where Re Y_{11} is measured.

We should be careful to note the significance of Eq. (10.24), for it identifies the cutoff frequency of our low-pass filter, where the gain of the distributed amplifier still remains near 8 dB. Our equation for the figure of merit, or gain–bandwidth [Eq. (10.5a)], on the other hand, identifies where our power gain for a conventional amplifier has dropped to 0 dB.

Careful design and fabrication of an m-derived low-pass distributed amplifier will provide excellent group delay to about 83% of f_c.

10.9 POWER MOSFETS AS MIXERS

As we saw in Fig. 4-8, the dynamic transfer characteristic of a power MOSFET would suggest that if we properly biased the power MOS-FETs, we should be able to achieve remarkable dynamic range, good sensitivity, and excellent resistivity to possible desenitization resulting from strong nearby signals. We might at first, question the feasibility of using short-channel MOSFETs as mixers simply because of their *linear* transfer characteristics; for two signals to "mix" to provide an intermediate frequency (*if*), we need mixing *products* which can only be generated using a nonlinear device. How, then, can we use a power MOSFET whose transfer characteristic is linear? As we may remember from our discussion in Chap. 4, all short-channel power MOSFETs have three quite distinct states in their transfer characteristics: a subthreshold region, a square-law region, and the constant transconductance region. The first two states are nonlinear and can be used for mixing; however, the nature of the nonlinearity plays an important part in the overall effectiveness of our mixer. In the mixing process within a FET, the value of ac drain current may be derived from a knowledge of the Taylor series power expansion:

$$i_d = g_m e_g + \frac{1}{2!}\left(\frac{\partial g_m}{\partial V_g}\right) e_g^2 + \frac{1}{3!}\left(\frac{\partial^2 g_m}{\partial V_g^2}\right) e_g^3 + \cdots \quad (10.25)$$

which, in turn, can be broken down into three terms:

Term	Output	Transfer Characteristic
1. $g_m e_g$	F_1, F_2	Linear
2. $\dfrac{1}{2!}\dfrac{\partial g_m}{\partial V_g} e_g^2$	$2F_1, 2F_2$ $F_1 \pm F_2$	Second-order and square-law
3. $\dfrac{1}{3!}\dfrac{\partial^2 g_m}{\partial V_g^2} e_g^3$	$3F_1, 3F_2$ $2F_1 \pm F_2$ $2F_2 \pm F_1$	Third-order

The first term is of no value to us and the third term offers us many harmonics but not the ones we need. The second term is our well-known square-law term, which is responsible for the generation of the important sum and difference signals necessary for mixing action. We also recognize that the square-law characteristic generates the fewest unwanted frequencies.

We can set the bias of our short-channel power MOSFET to emphasize literally any portion of Eq. (10.25). The constant-transconductance or linear region would be of little use, as there would be little or no mixing action. Similarly, the subthreshold region would not provide efficient mixing but would, instead, help mask what little mixing action we did have with many spurious frequencies.

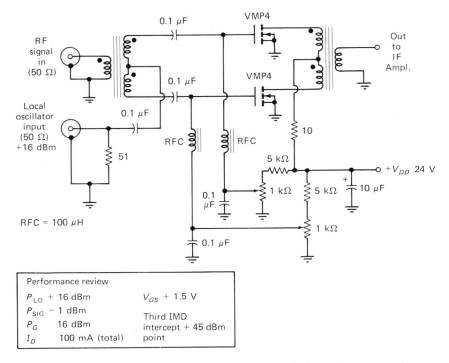

Figure 10-17 Single balanced high-level high-dynamic-range mixer using power MOSFETs. (Courtesy of *QST*, Jan. 1981.)

The balanced mixer, shown schematically in Fig. 10–17, provided conversion gain of 16 dB and a third-order *output* intercept point of +45 dBm when biased for a total drain current of 95 to 105 mA and driven with +16 dBm of local oscillator power. Decreasing the drain current saw a progressive decrease in conversion gain, whereas increasing our drain current saw a falloff in third-order intercept point performance, as we would expect.

There are several disadvantages in using power MOSFET balanced mixers in small-signal receivers. Although they provide a high dynamic range, as we have seen, we achieve this with substantial local oscillator power which we might find difficult to keep out of other critical areas. A second common problem with many FET mixers is their higher-than-normal noise figure. Whereas with a passive diode mixer we can easily obtain noise figures only slightly higher than their conversion loss, a power FET mixer will seldom offer less than 8 dB, even at moderate (3 to 30 MHz) frequencies. If we tried to compensate for this poor noise figure by adding preselection, the added gain necessary to reduce the second-stage noise figure would degrade our dynamic range.

Unlike the power bipolar transistor, which because of its resistance to burnout allows us to use spark-gap techniques for terminal protection, the power RF MOSFET frequently shows gross failures whenever the voltage spikes reached their drain–source breakdown voltage. As a consequence of this sensitivity to overvoltage destruction, we find that our terminal protection from *electromagnetic pulse* (EMP) is somewhat more complicated, as shown in Fig. 10–18. In this illustration we shunt the antenna with a Joslyn-type spark gap and series-couple through a delay line to a shunt bipolar limiter whose parasitic capacitance has

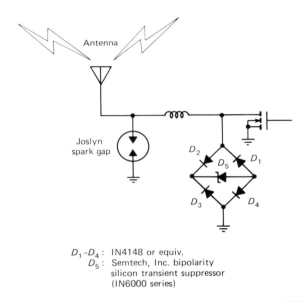

D_1-D_4: IN4148 or equiv.
D_5: Semtech, Inc. bipolarity silicon transient suppressor (IN6000 series)

Figure 10-18 Hybrid terminal protection for power MOSFETs, which allows them to operate safely in a potentially high energy environment.

been reduced by enclosing it within a diode bridge. Our purpose for the delay line, which generally can be nothing more than the series-resonant element of our tank circuit, is to provide sufficient delay to enable the incoming voltage spike to fire the spark gap before the diode limiter comes into play. The combination of spark gap and diode limiter becomes very effective in maintaining a maximum voltage spike leakage well below our critical voltage breakdown. In Fig. 10–19 we have a digitized display of a 30-ns 1.6-kV incoming pulse (a), effectively reduced to a manageable 90-V pulse only a few nanoseconds in width (b).

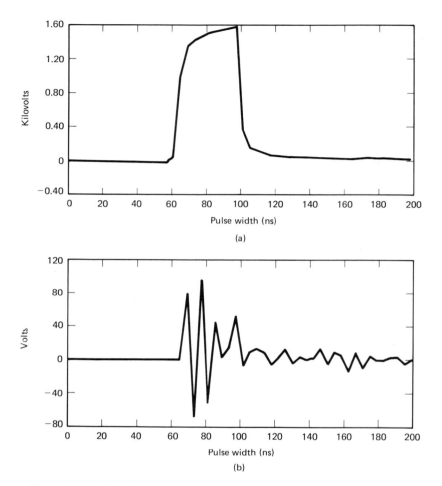

Figure 10-19 Digitized incoming 1.6-kV pulse entering the terminal protection circuit (Fig. 10-18). (a) The resulting output pulse. (b) Greatly attenuated voltage spike incapable of damaging the RF power MOSFET.

REFERENCES

DeMaw, M. F., *Ferromagnetic-Core Design and Application Handbook.* Englewood Cliffs, N.J.: Prentice-Hall, Inc., 1981.

Hardy, James K., *High Frequency Circuit Design.*   Reston, Va.: Reston Publishing Company, Inc., 1979.

Johnson, E. O., "Physical Limitations on Frequency and Power Parameters of Transistors," *RCA Review,* 26 (June 1965), 163–77.

Krauss, Herbert L., Charles W. Bostian, and Frederick H. Raab, *Solid State Radio Engineering.* New York: John Wiley & Sons, Inc., 1980.

LEIGHTON, LARRY, and ED OXNER, "VHF Power Amplifier Design Using VMOS Power FETs," *RF Design*, 3 (Jan. 1980).

PETROVIC, V., and WILLIAM GOSLING, "A High-Frequency Low-Cost VHF/AM Transmitter Using V-MOS Technology," University of Bath (United Kingdom), School of Electrical Engineering, Dec. 1977.

VAN DER ZIEL, ALDERT, *Solid State Physical Electronics*, 3rd ed. Englewood Cliffs, N.J.: Prentice-Hall, Inc., 1976.

eleven

Selecting the Right FET for the Right Job

11.1 INTRODUCTION

There should be no question in our mind that power FETs will proliferate and eventually we will be as comfortable with them as we were with the power bipolar transistors. The excitement of this new technology has spawned many vendors, and no longer are we concerned with second-sourcing problems; in fact our problem is even more basic. In this wrap-up chapter we try to resolve how to select the right power FET for the right job. If we can set this book down having arrived at that level of confidence, we should begin to feel comfortable and perhaps a little excited about the prospects of presenting a finished project using state-of-the-art technology.

Despite the burgeoning number of catchy trade names that proliferate, we recognize that there are only four basic geometries: the vertical *static induction transistor* (SIT), the *planar double-diffused MOSFET*, the *vertical double-diffused MOSFET* (VDMOS), and the *vertical V-groove MOSFET* (VMOS). Those catchy trade names identify the topologies. We must not lose sight of the importance of topology, as it is here that the vendor achieves the packing density to make possible his low ON-resistance as well as his manufacturing yields.

The latter resulting in either favorable or unfavorable factory costs, which directly affects our cost.

We have been inudated with power FETs from a score of vendors. There are n-channel and p-channel FETs. The SIT, as a member of the J-FET family, is known as a depletion-mode (Type A) power FET, whereas the power MOSFETs are enhancement-mode (Type C). Within the collection of power MOSFETs we have low-threshold logic-compatible devices and others that have high thresholds that cannot interface directly with logic. Fortunately, we can identify some of those once-unfamiliar terms on the data sheet, where beta is now replaced with transconductance and V_{SAT} with $R_{DS(on)}$. All our vendors unanimously agree that their products are free of secondary breakdown, but can they handle inductive loads safely if we use the FET as a power switch? We are also aware that to turn a power FET either ON or OFF requires first charging and then discharging the equivalent input capacitor. Does the data sheet offer clues to the drive requirements other than simply the gate–source voltage necessary to reach a finite level of drain current?

If we have a project in mind for possible inclusion of power FETs, these and other questions may need to be addressed so that we can develop a feeling of confidence.

11.2 SELECTING A SWITCHING FET

What we want in a switching transistor is unfortunately not quite what we get! The perfect transistor would be able to switch in zero time and exhibit no losses. In other words, we need a transistor of infinite speed and, when passing current, absolutely no voltage drop. Furthermore, we expect no current to flow when the switch is OFF. From our discussion in Chap. 4, we have to admit that although the power MOSFET is very fast, it certainly does not meet the standards expected of a perfect transistor. Yet we can be confident that by careful selection we should be able to find a power FET that will outperform practically any power bipolar transistor. Let us see if we can take a typical power MOSFET data sheet and from it find a convenient method whereby we can identify whether it would make a good switching transistor.

Before power FETs came all we had to do the job of switching was either thyristors or bipolar transistors. Although a bipolar transistor is current activated (we have got to inject minority carriers into its base) and the FET is voltage-activated (to get drain current to flow), we do recognize that when we are switching at high rates of speed, both the bipolar transistor and the FET are charge-controlled transistors. Consequently, we should expect that as far as their basic switching characteristics are concerned, they should have much in common. If that is true, we should be able to learn what we need to know about switching power FETs by examining switching transistor handbooks and comparing bipolar transistor data sheets.

From our earlier discussion, in particular Chap. 5, as well as from our examination of these sources of information, we recognize that to design an effective switch it is imperative that we understand the behavior of our transistor when it enters the saturation mode. If we were using a bipolar transistor we would also need to know its current gain, or beta, so we could set the drive. With our power MOSFET we know that the drain current will be set by the application of a known gate voltage (see Fig. 4-6). We need to determine our saturation voltage and how hard we need to drive the FET to achieve maximum efficiency. This we can do by following the example from a bipolar transistor switching manual: construct a graph of V_{DS} versus V_{GS} for constant I_D contours. Assuming that our power FET data sheet is deficient here, we need a data sheet that does provide detailed saturation characteristics at *low V_{DS} levels*. In Fig. 11-1 we have plotted the drain output static characteristics of a typical power MOSFET. From this plot we are now able to examine in some detail the cutoff region, the active region, and the saturation region. The latter we are able to define by the equation

$$V_{DS} = V_{GS} - V_{\text{TH}} \qquad\qquad (11.1)$$

Using a power MOSFET as a switch, we should have a fairly good idea of the switching current that we need to control. If we have a handle on I_D (the switching current), we are now able to use this figure to determine how much gate drive voltage is necessary, the futility of overdrive, and the absolute minimum saturation voltage we will achieve. Most of our knowledge gleaned from this one figure could have hardly been obtained using the conventional I_D-V_{DS} plot of output characteristics that we generally find on most data sheets. We should, be careful however, to note the operating temperature of these characteristics. Data provided at ambient (25°C) would represent the optimum performance, and we should be quick to take note that, for example, a switching drain current of 8 A driven to optimum saturation of 3.6 V (shown in the example, Fig. 11-1) may cause localized heating of the semiconductor chip and would, in turn, alter the performance somewhat.

Another look at the typical power MOSFET data sheet and we find that all capacitances are generally measured at gate voltages below threshold, and if any detailed information is given, it is to show the effect of variable drain–source voltage on capacity. This information is hardly suitable if we wish to determine the drive requirements for a

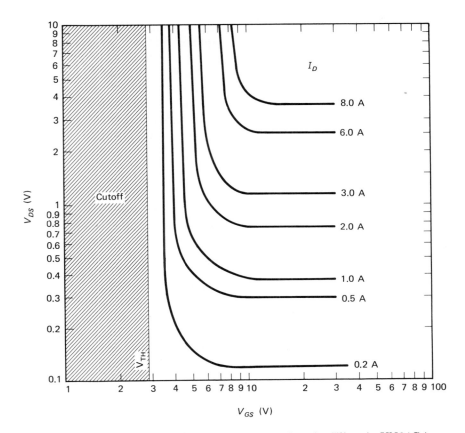

Figure 11-1 Drain output static characteristics of a Siliconix VN64GA VMOS power FET, showing the saturation voltage $V_{DS(on)}$ versus the gate drive voltage V_{GS} for various fixed drain currents I_D. Case temperature = 25°C. This plot identifies the exact amount of gate drive that is necessary for optimum $V_{DS(on)}$ for any known drain current (rms).

switching transistor. What we need are data that relate to the input capacity during turn-on. What we saw in Fig. 8-4 (for the same power MOSFET as has been plotted in Fig. 11-1) would be ideal but possibly too elaborate for most data sheets. A more reasonable effort might be to find a value for both C_{gs} and C_{gd} (the latter to determine the Miller effect) at some *operating* gate–source voltage which we could then take as a worst-case value to calculate our drive requirements and to predict our switching time. The fact that these capacitances do change is evident from Fig. 7-11.

Another very important specification that we must clarify is the suitability of our power FET's thermal properties. Here we may discover that the topology plays an important role. In our earlier study we examined the effect of packing density on current flow (Chap. 4 and Figs. 4-15 through 4-17); we can now relate the distribution of current flow to that of heat flow. Although beyond the management of the user of power FETs, we would find that soft-solder techniques of die attach offer superior thermal resistance than gold-eutectic die attach. Needless to say, our data sheet must offer a low thermal resistance, often identified as Θ_{JC}, with a numerical value provided in degrees per watt.

11.3 SELECTING A POWER FET FOR MOTOR CONTROL

If we assume that our mode of operation is PWM, then much of what we discussed in the preceding section is also applicable when selecting a power FET for a motor controller. Aside from a thorough understanding of the drive requirements (which is available from data similar to those shown in Fig. 11-1) and switching times—in particular how to switch OFF fast—we should be particularly concerned about the behavior of the FET when switching highly inductive loads, as would occur when using PWM for multiphase induction motors. A major concern is to select a power FET that can handle both the high standoff voltages and be able to switch substantial currents simultaneously and still have a SOA suitable to withstand *fault situations*. The most common fault would be a stalled motor, resulting in abnormally high current. A potentially more serious fault might be excessive slippage, due to overloading the motor, which would upset the power factor and result in both excessive current simultaneously with high voltage, which conceivably might exercise the SOA beyond its safe area unless we had taken care to choose a power FET with sufficient margin to account for such anomalies.

Motor controllers can be noisy, if not from the internal workings of the system, then quite likely from surroundings. Consequently, we may wish to consider high-threshold-power MOSFETs, which would be less sensitive to false triggering from random noise-voltage spikes.

11.4 SELECTING A FET FOR AUDIO AMPLIFIERS

In Chap. 7 we discovered that establishing the particular class of service (Class A, AB, or B) was dependent to a great extent on how well we could control the bias as the threshold voltage drifted downward with increasing temperature. One solution was to utilize a novel patented form of feedback (Fig. 7-18), but another potentially more practical solution might be to select a power FET with what is generally recognized as a *zero TC* sufficiently high to lie within the quiescent operating drain current of the FET. A zero temperature coefficient would ensure that over a reasonable temperature excursion, the fixed bias would maintain the correct class of service.

Perhaps the big question in selecting an audio power FET is: Should we use complementary pairs or design our amplifier using quasi-complementary n-channel FETs? If we choose complementary pairs, we must be aware that since the p-channel offers lower mobility than the n-channel FET, we must be careful to select a pair that offers a match of forward transconductance and drain current and accept the fact that the p-channel FET will have capacitances perhaps half again as high as the n-channel complement.

A question may arise as to the type of power FET we should select for our design; should we select a power MOSFET or a power J-FET such as the static induction transistor (SIT)? The output charac-teristics of the power MOSFET resemble those of the pentode, whereas those of the SIT are more closely identified with the triode. A more subtle difference between these two power FETs may be seen in their transfer characteristics. For the SIT, distortion tends to decrease as the magnitude of the load resistance increases, but for the MOSFET the opposite holds true. This follows from the fact that the transfer characteristics of the SIT become increasingly linear as the load resis-tance increases, whereas with the power MOSFET we see that the dynamic transfer characteristics are critically dependent upon the load resistance (or load line). The latter effect is clearly evident in Fig. 4–8. Being aware of the variableness that the loudspeaker's impedance presents to the output port of our amplifier, we might be tempted to select the SIT. But before we do we should be reminded not only of the lower amplification (μ) of the SIT, but that with this lower amplifi-cation, we will need higher gate drive, which conceivably might lead to dangerously high gate voltages (less negative tending toward a positive potential). As we saw in Fig. 2–18, this might convert our majority-carrier SIT into a minority-carrier bipolar transistor operating outside its SOA! The SIT, being a depletion-mode J-FET, also requires a gate bias of opposite polarity from that of the drain to ensure cutoff.

If our intention is to build a Class D switching audio amplifier, the requirements for the power FET are much the same as they are for selecting a switching FET.

11.5 SELECTING A LOGIC-COMPATIBLE POWER FET

We should not be surprised that this application restricts us to enchancement-mode MOSFETs. Of all logic, TTL is by far the most demanding, as its ON and OFF states are +2.4 V and +0.8 V, respectively. Unless we select an enhancement-mode power MOSFET with a threshold voltage only slightly above +0.8 V, we will sadly discover that the +2.4-V level will not provide sufficient turn-ON for many applications. Our problems, however, do not end with the selection of a low-threshold-power MOSFET. Since, as we learned earlier in this book, threshold voltage is temperature-dependent (a coefficient of approximately -5 mV/$^{\circ}$C for silicon), if we select a MOSFET with a threshold voltage of +1.0 V, we will need to maintain the chip temperature below 65°C. Consequently, we need to pay special attention to the thermal derating limits set by the data sheet. A more satisfactory solution is, of course, to use open-collector TTL, which allows us to use a higher-threshold MOSFET and, additionally, ensures a more forceful turn-ON.

11.6 SELECTING A POWER FET FOR AN ANALOG SWITCH

Because of the parasitic body–drain diode inherent in every power MOSFET, we became painfully aware in Chap. 9 that if we wished to take advantage of the low ON-resistance of power MOSFETs, we would need to use two back to back to effect a satisfactory analog switch. How this affects our selection is obvious: we have to find the lowest possible $R_{DS(on)}$ power MOSFET. We probably would have made this decision irrespective of the need for using two back to back; nonetheless, we must be aware of this added insertion loss.

As we mentioned in Chap. 9, we are severely limited as to the analog voltage peak-to-peak voltage swing simply because of the limited gate–source breakdown voltage. If we turn back to the basic analog switch circuit in Fig. 9-4, we see that in the ON state, the full analog signal potential sits between source and gate.

If our application will involve handling of appreciable analog signal current, we need to consider the voltage drop across our FET. A plot of V_{DS} versus V_{GS}, as shown in Fig. 11-1, is very illuminating, and from such a graph we can quickly determine if we need a heat sink. Generally, we would not for a switch, but it is good design to be absolutely positive, and Fig. 11-1 provides us a measure of assurance.

11.7 SELECTING A HIGH-FREQUENCY POWER FET

Even though there are only a few vendors offering a selection of high-frequency-packaged power FETs, we should not feel restricted in our attempt to make a proper selection; it all depends on our application. If we are considering large signal amplification, we want to stay clear of any power FETs with zener-gate protection, for reasons that we discussed in Chap. 4. Package style becomes important at high frequencies or when we are trying to achieve broadband performance. Incidentally, it is also very important when we are trying to switch very fast. A non-RF package, such as the TO-3, hurts us by introducing excessive series lead *inductance*, and since the MOSFET drain is common to the package, we often have problems maintaining a suitable heat sink while achieving a satisfactory high-frequency circuit. Over a moderately narrow band we can use such packages as the TO-3 and place beryllium oxide spacers between the package and its heat sink. A 0.065 in. (1.65-mm)-thick BeO spacer will add about 25 pF to the drain capacity of a TO-3 packaged power FET.

Of the many applications available for power FETs, high-frequency circuit design is sufficiently narrow that power FET data sheets addressing this market are generally quite conclusive. Aside from selecting the desired operating drain voltage and output power, we will find the greatest diversity in package styles. We should note one possibly important detail that might otherwise be overlooked in regard to our operating drain voltage. Be careful to note that the power FET's breakdown voltage is at least twice that of its rated operating drain voltage. We should be wary of power FETs with power ratings requiring, for example, 35 V, when their minimum guaranteed breakdown voltage is, for this example, say 65 V. That is satisfactory for continuous-wave operation but it would be disastrous for us if we wished to modulate the carrier to 100%.

Index

A

Abrupt junction diode, 102
Advantages of using power FETs:
  in audio, 183–84, 212
  in high-frequency applications,
    264–66
  in logic control, 240
  in motor control, 176–77
  in power supplies and regulators,
    154–55
  as a switch, 260–61
Amplifiers:
  audio, 180–212
    class of service:
      A, 195–97
      AB, 186, 195–96, 201–2,
        208, 211
      B, 195–98
      D, 203–8
      G, 208, 211–12
    cut-off frequency, 195
    design goals, 188–90
    distortion, causes of, 183–84
      (*see also* Distortion)
    modeling, 114–18
    problems defined, 180–81, 185–86
    push-pull, 195
    quasi-complementary, 186
    selecting the FET, 300–301
    single ended, 190–92
    source follower, 190
    Static Induction Transistor,
      186, 195
  high frequency, 262–95
    class of service:
      A, 278, 280, 285
      AB, 281–83
      B, 278, 285
      C, 195–96, 277–79, 285–86
      D, 286–87
      E, 287–88
    distributed, 287–90
    figure of merit, 22, 93–94, 139,
      273
    linear, 283–86
    power gain, 276
    push pull, 278–80
    stability, 279, 282–83
    systematizing design, 275–80